What Pest Control Companies Don't Want You To Know

Scripture reference is taken from the King James version of the Holy Bible.

ISBN: 1-878898-13-2
LCCN: 97-076882

99 98 97 96   95   10 9 8 7 6
Printed in the United States of America

CAP BOOKS
P.O. Box 531403
Forestville, MD 20753

Acknowledgments

Dr. Avaneda Hobbs sends special thanks to some of the most fantastic people God could give to anyone's life. First, lots of love, hugs and kisses to my incredible mom, Mrs. Viola Hobbs, my fantastic sister, Iris Lynette Hobbs-Dixon and son, Tyrone "silky smooth singing" Dixon. To Frederick ("Ricky") and his wife EmmaLisa and family. To Harold ("Sonny") and his wife Cheryl and family. I also send special thanks and love to the entire family on my mom's (the "Walker, Shepherd and Brock families") and my dad's side (the "Hobbs and Patterson families"). Incredibly gigantic thank-yous and kisses to my big brother, the one and only Demond Wilson. Gigantic kisses and hugs to Nellie "Aunt Pal" Ceasar, Attorney Kimberly Lewis, Earnest and Robin Workman and MSgt. Larry Harris. To my two godchildren, Whitney and Wesley, I love you.

Roosevelt McNeil wishes to thank Pastors Tony and Cynthia Brazelton. Also, special thanks go to his parents, Roosevelt Sr. and Bettie L. McNeil for trusting God in raising me and giving me a respect for the elderly. My personal thanks are to two very special people, my brother and sister in the Lord, Eric Ingram and Dallah Herman. May the Lord continue to keep you in His presence and bless you with all prosperity. My personal and special thanks to Edward D. Lee for his love, patience and caring for me. Ed, I love you man.

Roosevelt would also like to thank Dr. Avaneda Hobbs for taking an idea submitted to her last year and turning it into this wonderful reference book. Dr. Hobbs, you have been incredibly wonderful to me. Thanks for the countless hours of support, love, pouring your "professional wisdom" into me, and helping me to prepare for the "worlds' educational stage." May God bless you richly far and beyond what you could ever ask for.

This book is dedicated to my loving children whom God has entrusted into my care. They are: Roosevelt III, John Anthony, Toya, Nekia Rose and Serina Rebecca and (little one) Troy.

Foreword

This book is a powerful tool that the Lord has provided to the nation. It is designed to help you gain insight into pests, pest control management and the pest control industry.

Micah 6:8 says, "He hath shewed thee, O man, what is good; and what doth the Lord require of thee, but to do justly, and to love mercy, and to walk humbly with thy God?" Simply put, do what is right! How many times has this statement simply been forgotten . . . do what is right? This book was written because of the integrity in the heart of the authors. The primary goal is to emphasize what has been stated . . . do what is right!

TONY BRAZELTON, PASTOR
Victory Christian Ministries
Clinton, Maryland

Preface

Every one of us has been confronted with the problems of pests, from cockroaches to ants and from mice to rats. We have spent literally hundreds and thousands of dollars on pesticides, chemicals of every kind, and on hiring pest control companies. Still, we get frustrated and disgusted every winter when mice and rats invade our homes and businesses. We become downright angry during spring and summer when insects of all kinds (i.e., cockroaches, crickets, spiders, fleas, ticks, mosquitoes, bees, moths, beetles, ants, etc.) invade our businesses, homes, lawns and gardens. They can cause literally hundreds of thousands of dollars in damages to our property and in costly repairs.

Learning how to manage and take control of our lives are not easy. We struggle with the day to day trials and tribulations of life and seek a solution and a way out. There are solutions and there are ways to get out from under the burdens of life that we face. I'll deal with most of those at another time and place.

The authors of this book, "What The Pest Control Companies Don't Want You To Know," have given all of us some powerful information. This information is necessary for us to manage the problems of controlling the various pests that invade our homes, lawns, gardens and businesses. Because of this book, we can now manage the pests and rodent problems in our lives.

Presented in this marvelous book are eight chapters and a powerful appendix. Chapter One gives us such great information as what the National Pest Control Association defines as the most destructive common household pests. Chapter Two, titled "The Trap," tells us what happens and what to do when an inspector comes out to entice you into a maintenance contract. Chapter Three informs the reader about what insects come, when they come out, and what we should do during those seasons. Chapter Four walks us through a process on how to conduct your own household pest inspections. Chapter Five provides a brief narratives, along with the pictures of each, of the most common household pests. The narrative includes information on their color, life cycle, home and nesting sites. Chapter Six offers recommendations on what pesticides to use and how they work. Chapter Seven provides information on the common household insecticides and their uses. Chapter Eight instructs the reader on the cautionary procedures to use before signing pest control company contracts. The appendix provides a list of retail outlets where pesticides can be purchased, list of state and government agencies, and a listing of resources on the Internet. Needless to say, a wealth of information is provided in this book. You will refer to it, time and time again!

"What Pest Control Companies Don't Want You to Know" should be on every book shelf and in every household in America and around the world. It is one of the best, if not the best, resources ever put together to help you manage and control the pest problems in your life. Applying the knowledge and recommendations of this book will save you thousands of dollars and help you have a pest-free life, at least about insects and rodents!

As you began your journey to learn how to manage and control your life better, including this book on your list of places to visit would be wise as you stop along the road of life. You won't regret the stop you made and the time you invested in reading this book. You'll wonder how you ever managed your life without it!

LARRY DARNELL HARRIS, M.SGT.
Virginia Air National Guard
Norfolk, Virginia

Table Of Contents

Introduction

We all know what pests are -- animals or insects that can influence our lives positively or negatively. Despite all that you know or have heard about pest control, strip yourself of it all. We are going to delve into the truth.

The *National Pest Control Association* (NPCA) says that "pests can transmit as many as fifteen major disease-causing organisms. Some of those diseases are the Plague, Rocky Mountain Spotted Fever, Hauntorus Virus and Lyme Disease. The NPCA lists in their Pest Fact Sheet that "seven to eight percent of the general population is allergic to cockroaches." "Cockroach suppression and eradication" they say, "are vital to health care facilities, homes and wherever food is prepared and served. Cockroaches contaminate food and transmit disease carrying organisms, including staphylo-cocci, strep, coliform molds, salmonella, yeasts and clostridia."

Diseases Caused by Pests
★ The Plague
★ Rocky Mountain Spotted Fever
★ Hantavirus
★ Lyme Disease
★ Staphylococci
★ Strep
★ Coliform Molds
★ Salmonella
★ Yeasts
★ Clostridia
★ Rat-bite Fever
★ Salmonella
★ Trichinosis
★ Murine Typhus
★ Leptospirosis

Further, the NPCA says that "rats bite more than 45,000 people each year in the United States. They transmit many diseases, including rat-bite fever, salmonella, trichinosis, murine typhus, the plague and leptospirosis." Each year, according the NPCA, pests are responsible for $3 billion in damage. For example, the NPCA says that each of the 200 million rats in America can cause as much as $1,000 in damage.

The NPCA also offers other alarming statistics on damages caused by pests. In 1981, their written Pest Fact

Sheet showed that termites caused an estimated $735 million in damages. In 1983, the *Entomological Society of America* reported that in the nine southeastern states where they conducted studies, an estimated $579 million in termite damage was caused. Recent estimates put the total cost of termite damage and control between $1.2 billion and $2 billion.

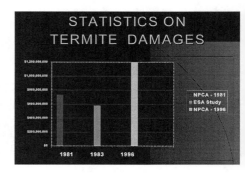

Let's go a little further and define pest control. The New Webster dictionary defines control as to govern, exercise control over, to restrain or to regulate. Pest control is therefore defined as the means of governing, controlling, regulating or restraining animals categorized as pests.

The pest control industry is one of the largest industries in North America. In 1996, the NPCA said that the pest control industry had estimated gross revenues of $4.6 billion. According to 1996 estimates by the NPCA, the pest control industry in the United States is composed of 18,000 companies. These companies employ nearly 85,000 service technicians.

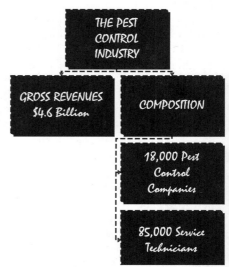

The NPCA also says that each year, the pest control industry provides pest control services to 12 million homes and 288,000 retail food outlets. In addition, they provide services to 480,000 commercial restaurants and kitchens, and 66,000 hotels and motels.

Usually, a person cannot sue a pest control company. Why? Pest control companies never say that they will get rid of your pests. That is why they put the customer on a pest control maintenance program.

On a one year agreement or contract, if a pest control company comes out often, it will make them more money. It is their objective to prolong the service . . . and they do. Ask yourself this question -- does it take one year for a company to get rid of a pest?

I first became interested in pest control and the pest control industry in 1990. At that time, I was working for a very well known commercialized pest control company. During this time, I increased my knowledge of the pest control industry and how to control and elimi-nate pests. This included treating homes, offi-ces, restaurants and various buildings for them.

Experience at my second job, with another well known pest control company, opened my eyes to the need for each consumer to become educated in three specific areas about the pest control industry. First, they need to know the best ways of receiving the highest quality service from pest control companies. Second, they need to know how to maintain a pest free environment. Third, they need to know how to save money when buying products for pests and in using pest control companies.

Why write this book? It touched my heart to see elderly consumers with a need for this service. Most were on a fixed income and unable to pay for a "one year" contract offered by many pest control companies. In my experi-ences with these pest control companies, I was told not to do "one time" treatments for the customer. This was because the company wanted them to sign a one year contract, which was totally unnecessary in most cases.

THE PEST CONTROL INDUSTRY

TYPES OF CUSTOMERS

12 Million Homes

288,000 Retail Food Outlets

480,000 Commercial Restaurants and Kitchens

66,000 Hotels and Motels

These consumers had a need, yet they were unable to get good quality service for a reasonable price. Some customers were willing to pay, even able to pay, yet they were unable to receive the appropriate treatments they really needed. Why? It was not helping the consumer that counted. It was all about making a profit for the company.

After getting substantial knowledge and awareness of the biology of pests and how they live, eat and breed, this information is now shared with the readers of this book. The purpose of this book is to free individuals, corporations, restaurants, churches and other establishments from the pest control trap.

The Trap

There are probably many questions on the minds of consumers today about pest control. Many consumers believe 99% of what they are told by pest control companies. For instance, Mr. and Mrs. McNaps have ants all around their kitchen. Mr. McNaps tried to get rid of them himself, but failed. Mrs. McNaps calls an exterminating company. The exterminating company sends out a sales representative to inspect the problem. From this point, two things usually happen.

Example 1 -   An inspector comes out, looks in one or more places within five minutes or so. He picks up an ant and says "Mrs. McNaps, you do have a problem. We can help you but we would have to put you on a maintenance program." ***Listen to the sales representative's line***. "We can 'cover' you not only for ants, but for spiders, cockroaches, fleas, ticks, rats, mice, crickets, etc. All you would have to pay is $35.00 dollars or more a month. The start up fee is only $96.00, which includes the first month's service."

Example 2 -   An inspector comes out and goes straight to the kitchen. What you have been offered concerning your pest problem, is a contract. Yet, you simply wanted to get rid of the ants. Did anyone say "Mrs. McNaps you can go to the store and buy ant baits or sprays that would get rid of the ants?" Did anyone say that she could use lemon juice or vinegar to get rid of the ants? Further, did they say that it would be better and cheaper if the McNaps' got rid of the ants themselves before calling the exterminator? No! Has any company suggested how and where to get these products?

There are special warehouses that carry most of the same products that exterminating companies use. These are nonrestricted products for customers to purchase. Others are restricted products for licensed persons to purchase and use. The only requirement for customers is to *"ALWAYS READ THE LABEL BEFORE USING ANY OF THESE PRODUCTS."*

Most pest control companies are not geared to come out and get rid of your problem. Their primary concern is to get you to sign a one year contract for a maintenance program. Their reasoning is, if they came out and got rid of your problem in one or two treatments they would make very little money. If pest control companies use the same chemicals to get rid of your problem in one year, then what is stopping them from getting rid of them in one or two treatments. They do not want to spend the time because they are charging by the hour. Their concern is not essentially for your comfort or your need, nor to get rid of your pest problem. For example:

A company sends out an inspector. They give you a free inspection and then write up a proposal telling you how much the service costs. If you sign the "Agreement" or contract, the salesman gets 15% of what he, or she sells. A technician gets 10-15% and the company receives the

remainder. The customer is paying the salesperson for the company, the inspector and the technician. Thus, the customer is not paying for an expedient method of getting rid of the pests.

Sometimes there may be a need for several treatments. However, a contract that locks the consumer into a high priced long term agreement is not in the interest of the consumer, but the company.

There is no doubt that some areas may need more service than others, but the consumer should not be exploited. Pest control companies know that these consumers have a need and are willing to do whatever it takes to get rid of their pest problem. Consumers in this category are the primary targets.

What is being stated here, is that profit is the bottom line not the consumer. It's about profit. It's not about educating or helping the consumer. For the consumer, it is not about helping you to go a cheaper way and simultaneously receive quality service.

The pest control industry has become a billion-dollar business. They have made their profits off the needs of consumers, like yourself. I cannot say that all pest control companies are bottom line profit oriented, but based on my experiences, most of them are profit oriented.

Seasonal Exterminations

Most pests cannot live outdoors during the winter months. Monthly services are not normally needed for pest control on the outside, during the winter months. If pest control services are required, only one to three treatments, at most, are needed. For example, if the consumer has pets who have had fleas, it is good to have the inside of the home treated. This should be done once or twice during the winter months. Fleas live in a larva stage, which is almost like a cocoon and remains buried in the carpet of the home for up to eight months.

During the winter once the temperature has reached 75°-80° indoors, or the carpet has been vibrated, fleas will jump out. It is easier to kill adult fleas than the flea larva because they are protected by the cocoon. Once your carpet has been treated for fleas, it is a good practice to vacuum twice a day. Vacuum the carpet for at least seven days. It is imperative that the vacuum bags are thrown away each time the carpet is cleaned. After the seventh day, vacuum only once or twice a day. This will help pick up any adult fleas left behind.

There are other pests that the consumer would really want to treat during the winter months. They are cockroaches, rats, mice and ants.

Let's go over this again. Why does the customer sign

> ### FREE FLEA
> #### Winter Season Checklist
>
> ✔ Have your carpet professionally treated.
>
> ✔ Vacuum twice a day after treatment.
>
> ✔ Vacuum the carpet for at least seven days.
>
> ✔ Throw vacuum bags away each time carpet is cleaned.
>
> ✔ After the seventh day, vacuum only once or twice a day.

a one year agreement? Why does the customer pay for twelve months of service? Has the customer asked him or herself why it takes the pest control company twelve months to get rid of ants? Or for that matter, why does it take twelve months to get rid of spiders, crickets, rats and mice? Why?

Again, why a year contract? Why not two or three years? Why not six months or three months? Why not even a month? If the pest control company and the consumer have the same chemicals to get rid of pest problems in a year, what stops them from getting rid of pests in one to two treatments? If it is the same chemical, then the only answer to the delay is TIME. Time is the only element left for consideration.

The average service technician spends time building his knowledge base on pest control in five specific areas. They are: (1) getting to know the biology of pests, (2) finding out where they live, (3) learning how they breed, (4) knowing what kind of pests they are, and (5) knowing how long it takes to destroy their nests.

For years the consumer has been paying for twelve months of pest control services. Further, they have been getting unsatisfactory results. Some cases have been good, however, the majority have received unsatisfactory results. Even if the consumer has had some success with their pest control company, most have probably paid two to three times the normal amount. Most, but not all, pest control companies are out to make a profit.

There is nothing wrong with making a profit. We all would like to make a profit from services rendered. This could be either on a job or a privately owned

Understanding Pests

1 Get to know the biology of pests.
2 Find out where they live.
3 Learn how they breed.
4 Know what kind of pests they are.
5 Know how long it takes to destroy their nests.

business. What is being expressed here is, no one should make a profit at the literal "expense" of someone else. A person should make an honest profit from services rendered.

If a company promises to get rid of the consumer's pest problem within a year, then this should be performed . . . especially after a year has ended. The most important thing for the consumer to receive is a thorough inspection and to get the right type of treatment. In Chapters 4 and 6, we'll discuss inspection and treatment of household pests and consumer concerns.

Conducting Household Pest Inspections

To do a thorough inspection in and around your home, or business, the consumer must first understand what is an inspection. Webster's New World Dictionary says that an inspection means to look at carefully or to examine officially.

No one knows a consumer's home like they do. The consumer sleeps, eats, and lives in their home year after year. The consumer is familiar with what time of the year ants, mice, mosquitoes and other pests try to invade their property. The consumer is the inspector. Therefore, the consumer is qualified to do inspections. The consumer does not need a license to do an inspection in his or her home or business. They just need to get the job done.

Since we have defined inspection, let's talk about the type of equipment that is needed to do a thorough and efficient inspection. A flashlight and a probing tool are needed.

The flashlight is a very important tool. It helps one to see areas that the naked eye may miss under normal circumstances. These include areas such as the inside walls, under shelves and in dark closets.

Most pests live and breed in dark areas. For example, German cockroaches do not crawl directly into a hole. They back into a hole, leaving their antennas sticking out. German cockroaches also use this to detect pesticides, insecticides, chemicals or any other type of danger that may be present.

A probing tool is another good instrument. One effective type of probing tool is a flat head screwdriver. Another is a long ice pick. This will aid the consumer in getting to areas where the hand cannot reach into or where the consumer may not want to stick his or her hand. The screw driver or ice pick can also be used to inspect for termites and carpenter ants.

Identifying Common Household Pests

There are several common household pests in America. They include cockroaches, pantry moths, fleas, rats and mice, mosquitoes, slugs and snails, bees, termites, spiders and beneficial pests. Each is categorized and discussed in subsequent subheadings in this chapter.

Cockroaches

With the right amount of time spent during the treatment, cockroaches can be eliminated. Yes! It's true. Do not be fooled by anyone on this matter. There are also some easy steps that can be taken to get rid of this problem.

One step to eliminate cockroaches is to cork around kitchen sinks, counter tops and windows. The consumer should also cork around doors, kitchen pipes, behind stoves and underneath cabinets that lead into walls. German cockroaches live in these areas, because they are dark and have deep voids with warmth and moisture. German cockroaches need three things to survive -- water, food and shelter. There are essentially four breeds of cockroaches.

1   Cockroaches that are seen running around in kitchens or bathrooms when the lights are turned on are *German Cockroaches*.

2   Cockroaches that are dark reddish brown, large and sometimes fly (they cannot really fly, they glide from a higher position to a lower position). These are called *American Cockroaches*.

3   Dark black cockroaches that are usually outside, in sewers, and basements are called *Oriental Cockroaches*.

4   Cockroaches that are found near furniture and behind bedroom dressers are called *Brown-banded Cockroaches*.

German and Brown-banded cockroaches are almost alike in appearance. However, they are different. Both can be found in bedrooms. Brown-banded cockroaches are very rarely found in kitchens. German cockroaches travel 12-15 feet from their nesting areas. The reasons for seeing cockroaches in the daytime are overpopulation, being disturbed or that an insecticide has been applied.

Corking will help eliminate 90% of pest entry ways into your home, office or building space. Here are other tips that will help prevent cockroaches from living in your home or building space.

♦   Do not use paper bags in your home or building space. Use plastic bags. German cockroaches and silverfish like the glue and paste that hold paper bags together.

♦   Replace old and worn seals at the bottom of entry and exit doors and garage doors. Make sure that they are snug and tightly fitted. This will also help to keep out rats, mice and other pests.

♦ Make sure all windows and door screens have a snug fit. Replace screens that have holes in them and seal all cracks around windows.

♦ Replace food items kept in bags or boxes in kitchen cabinets and pantries. Store these food items in your refrigerator or in plastic type storage containers with air tight seals. This will help eliminate a food source for cockroaches and pantry moths.

♦ Wash kitchen floors, counter tops and occasionally dishes with bleach or with a product with bleach in it. This will help eliminate germs and bacteria that may be left in unseen areas. Doing this will also help to eliminate 99% of pest food and shelter.

♦ To help eliminate pests such as beetles, crickets, fleas and other outdoor pests, keep grass cut short and neatly trimmed. Keep the trees and bushes trimmed so that they do not touch the house's walls and windows. Pests, then, will be unable to climb into your homes and building spaces.

♦ Keep gutters clean and free from leaves. Do not let leaves build up in them. Moisture will be eliminated and not allowed to build.

- Keep garbage areas clean, with their tops tightly fitted.

Pantry Moths

The same treatments, sanitation programs, and food storage methods used for cockroaches should be used for pantry moths.

Fleas

In this section, we will discuss how to treat your home or building space for fleas. We will cover how to keep fleas out of your home or building space, once they have been eliminated. Indoor fleas have limited places where they can live comfortably. They are in carpets, furniture, and animal bedding or on the animal itself.

Fleas cannot live on humans like they can on pets because we take baths and showers daily. However, they will bite and some bites can be very painful. Some may experience an allergic reaction. Fleas are carriers of diseases. Therefore, a thorough treatment must be done.

Areas for treatment of fleas are pillows on couches, chairs and rugs. Wood floors and kitchen linoleum must be treated very cautiously. Treat underneath beds and bedding if needed. Bedding should be washed in very hot water. All pet bedding must be treated very thoroughly. After treatment has been done, vacuuming will help pick up the adult fleas. This will also help to keep them from reproducing more eggs. However, there may be larva left in the carpet. (Bombs do not affect the larva.)

The larva cannot be affected by the pesticide until they hatch and then touch the treated carpet. The larvae are fleas protected inside a cocoon like shell. They can live in cocoons for up to eight months inside the carpet.

If the consumer has pets, it is good to have inside flea treatments at least twice a year. If the consumer does not have pets, a one-year treatment is sufficient. This will not only work for fleas but will also work for carpet beetles and other home invading pests. Keeping dog and cat bedding areas clean and treated will help to eliminate breeding and hiding places for fleas.

For outdoor flea treatments, the grass must be cut short and kept neatly trimmed. What is not told is that fleas only live in shaded areas of the lawn, and under trees and bushes. Fleas will not live in sunny hot areas of the lawn because they will dry up. Keeping your lawn cut short will help eliminate shaded areas for fleas. Fleas do not actually live in dirt, however, some ticks do.

Rats and Mice

In the extermination/treatment of rats and mice, make sure holes that allow entrance into homes, and building spaces are tightly sealed. A foam sealant or corking material should be used. Make sure that air conditioners, heat pumps and cable cords are sealed. Dryer hoses also need to have a sealant around them. Make sure all drains are draining properly. This will keep water from flowing toward your home or building space. Likewise, it will keep water from seeping through the foundation into the basement.

Remember all these conditions are favorable to rats and mice. Rats can get through a hole the size of a quarter. Mice can get through a hole the size of a dime. Once these pests have gained entrance, they will live in voids inside of walls and in furniture. Roof rats can live in the attic.

When rat and mice's droppings are found, be very careful when cleaning them up. Rodent droppings carry a virus called the *"Hauntorus Virus."* When coming in contact with rodent droppings wear a dusk mask and gloves. (These masks can be purchased at any hardware store.) This will keep you from breathing in the virus.

When a consumer has rats or mice, there is one thing that the naked human eye cannot notice or see. That is rodent urine. The consumer needs a special black light to see rodent urine. This can be purchased at any wholesale or hardware store and are normally used by professionals to see the urine trails of rats and mice.

If rats or mice have been in the kitchen areas, such as cabinets and counter tops, wiping these areas down daily with bleach is very important. Bleach is a very good disinfectant and a germ and bacteria's killer. This is not to say that all other disinfectants do not work. However, bleach is the best product for the job.

Mosquitoes

Mosquitoes are the most dangerous insect pest to mankind. There are many mosquitoes and most of them are harmless, except for the pain or discomfort of their bite. Few are carriers of diseases which causes death to many people and the serious illness of many each year. Mosquitoes are believed to be the only agents that carry and transmit malaria, yellow fever, dengue fever, and filariasis to man. They are believed to be the leading agents in transmitting several forms of *"Virus Encephalitis."*

The best way to treat mosquitoes is to have a mosquito repellent plant. Approximately 98% of the people have never heard of a mosquito plant. Most garden nurseries carry these plants. They range in price from $3.00 to $5.00 per plant. Some may be more expensive, depending on the store where the plant is purchased.

The mosquito plant is from the geranium family. It has a unique fragrance of the citronella plant. It is a scent people love, but mosquitoes hate. Mosquitoes will not come within 100 yards of these plants.

Slugs and Snails

Slugs and snails kill and eat most plants and vegetables. They are agricultural pests and have to be treated by a special licensed person. However, the consumer can treat these pests with natural solutions without using pesticides. The simple old-fashioned way is to sprinkle salt or garlic around your garden and flower beds. Or, the consumer can place a bowl of beer where slugs and snails are active.

Ants

Ants are known for their persistent set of attitudes in which they live. They are a benefit to man in many ways. First, by turning the soil in their living space, they circulate air through it. Second, they destroy termites, cockroaches, and other pests that feed on farm crops.

Ants are considered pests because they invade homes, commercial buildings and warehouses. They seek to store food and in the process destroys growing plants. These creatures can be found in bathrooms, plumbing fixtures, foundations, sidewalks, porches, baseboards, electrical outlets and kitchens. They are especially fond of sugar and grease (because of the fat in grease).

Ants can damage wood, pollute food, eat holes in fabrics, steal seeds from plants, gardens and of lower beds. They are also carriers of disease.

To track food, ants use their senses. They have a very well developed sense of taste that enables them to distinguish between things that are sour, sweet, bitter and salty. Also, they have a keen sense of touch and use their antennas to smell, taste and touch. There are probably about 10,000 or more ant species in the world.

Termites

Termite is a common name of a group of insects that live in communities like ants. Because of some likeness to ants and living in colonies like ants, they were given another name -- white ants. Termites do have habits like ants and live together and are small like ants, but are of a different order of insects. They are more related to cockroaches and grasshoppers.

There is probably 3,000 or more termites in the world, but about 2,000 are known. About 30 or more species are known to live in North America. In the United States termites fall into three groups. These groups of termites are called subterranean, damp-wood and dry-wood termites.

The likeness is that they have feelers (antennas),

Common Household Pests

X Cockroaches
X Pantry Moths
X Fleas
X Rats and Mice
X Mosquitoes
X Slugs and Snails
X Ants
X Termites
X Spiders
X Beneficial Pests

mouth part likeness, thick waists, and primitive wings in a resemblance of that of cockroaches, than that of an ant. Ants have thin waists, elbowed feelers (antennas) and wings for a particular purpose. They are one of the oldest groups of insects in the world, having been in existence for over one million years.

All termite classes contain both sexes. The king lives as long as the queen. There are three basic classes in most termite colonies. Reproductive, workers and soldiers. In some species, there is only one class of reproduction. It consists of a queen and a king who are fully developed into a perfect male and female. They have fully developed wings, eyes and moderately hard bodies. The queen termite is the largest.

When the *Queen* is swollen with eggs, she can reach up to one inch long. When swollen, she looks white except an appearance of broken dark bands around her body. The queen lays eggs at the rate of several thousand a day. When the eggs are laid, the workers carry the eggs away and place them in very special constructed cells. The workers then feed and care for the young, which are called larvae, as they hatch from the eggs.

The *Workers* and *Soldiers* are usually sterile. They have no wings and are blind and small. Their color is pale or milky white, with soft bodies. They are about ¼ of an inch long. The workers are a very large number in the colony. Workers do all of the work. Specifically, they make tunnels, feed and care for the colony and do all the searching for food and water.

Soldier termites also have no wings and are blind. However, they are larger than the workers. In some species, soldiers do have very small under developed wings. They have large hard heads, strong powerful jaws for biting, snapping and grabbing. Their legs are also stronger, but their bodies are soft and weak. Some species release sticky or irritating liquid to repel off their enemies. Termites and ants are enemies. The only duty of soldiers is to guard and defend the colony against such insects like ants.

Most termites eat and digest wood, wood products, paper and many other materials containing cellulose (a fibrous carbohydrate) substance found in most plants. With the aid of a microscopic single-celled organism, called Protozoan, this helps their bodies digest cellulose. Then, it is broken down into carbohydrates that are easily absorbed by the termites' digestive system. There are some termites that eat grass and other herb like plants without wooden stems. Others refine or improve fungi for food.

Termite damage can be very costly if not discovered in the early stage. Termites on the other hand are beneficial to man. How? And in what way? They produce an abundance of fertilizer. By loosening the soil, they make the soil more absorbent of water. Also, termites turn other plant materials into organic matter.

Spiders

Spiders are eight legged animals that make silk (webs). The reason for them spinning webs is to catch insects. Insects are food for spiders. Whether small, or large, insects cannot escape the spider's web. All spiders spin silk but all spiders do not make webs.

When the insect comes near, the spider swings the line catching (sticking) the drop of silk to the insect therefore trapping it. Not only do spiders use their silk for catching food, but also to escape from danger. When you see a spider hanging down on a string of silk and danger is near, the spider will drop down on this line and hide until danger passes. The string of web is called a spider's lifeline or the drag line.

All spiders have fangs. Many kinds of spiders have poison glands. The bite of spiders can kill insects and some small animals, but very few spider bites are harmful to man. There are a few kinds of spiders in North America, which have bites harmful to man and can be fatal. Only the female widow spiders are known to bite humans.

Many people have a fear of spiders not understanding that the spiders will only bite humans when they are frightened, hurt or threatened. Spiders are beneficial to man because they will eat harmful insects. They eat grasshoppers, locusts and other insects, which destroy farm crops. They will eat flies, mosquitoes and other insects that carry diseases. There are some spiders, the larger ones, which will eat animals like mice, birds, lizards, frogs and fish. Spiders will even eat each other. Many female spiders are larger and stronger than male spiders, and will often eat the male.

Spiders will live any where and in several places where they can find food. Some of these include fields, woods, swamps, caves, deserts, inside houses, outside houses and in barns. Others include sheds, in cars, in buildings, outside buildings, in and outside windows, screens, doors and walls. There are probably 30,000 or more known spiders in the world, from the size of the head of a pin to the size of a man's hand. A South American tarantula can measure up to 10 inches long or more with its legs extended.

Spiders are not considered insects. Spiders are classified as "arachnids." Arachnids are groups of animals with eight legs. They include scorpions, granddaddy long leg spiders, mites and ticks. Ants, bees, beetles and many other insects have six legs. Spiders do not have wings and antennas like many other insects. Spiders are classified by scientists, called entomologists, in two ways as true spiders or tarantulas. Spiders have no bones. Their tough skin serves as an outer skeleton. Many spiders are covered with hair, humps and spines. Spiders have two body sections. They are the cephalothorax (head joined to the thorax) and the chest and abdomen (the hind part). They have a thin waist called the pedicel that attaches other parts called appendages.

Beneficial Pests

There are some pests that work for us. Did you know lady bugs, grand daddy long leg spiders and praying mantises' are beneficial pests that work for us? Why? These pests eat other pests that are threatening to us. They prey on mosquitoes, gnats and moths. These pests destroy our gardens, flowers, vegetable plants and invade our homes.

Did you also know that the grand daddy long leg spider is one of the most poisonous spiders in the world? They carry venom. But do not worry, they cannot pierce the human skin.

NOTE: On the following pages are pictorial narratives of the various kinds of common household pests.

Cacknoches

American Cockroach

Color - Light brown with reddish - brown wings, with a yellow perimeter on the back, between its head and wing.

Description - Eats or disfigures food, book bindings, wall paper and fabrics. Gives off a very bad odor. Pollutes food and is a possible carrier of diseases. They are poor flyers, but can glide considerable distances from high spots. They will feed on all foods, including decayed organic matter, paper, paste, starch, syrup and sweets.

Life Cycle - Adults can survive two to three months without food, but only one month without water. Others can live one to two years or more depending upon climate conditions.

Home (Nesting Site) - Will live in any room in a building, or beneath a building and outdoors. Prefers damp, warm places. They are common in food establishments. They also live in sewers, water meter pits and under rocks.

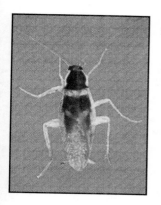

Brown-banded Cockroach

Color - Yellow to reddish brown with two cross bands on wings.

Description - The female is short and broad with wings barely reaching the end of its body. The male is slender and long with wings extending beyond the end of its body. Prefers to hang around furniture. Eats or disfigures foods, book bindings and fabrics. Gives off a very bad odor. Pollutes food and is a possible carrier of diseases. Its favorite foods are starchy materials, including wallpaper paste.

Life Cycle - Six months or two generations per year.

Home (Nesting Site) - Furniture and ceiling moldings. Prefers areas that are 80 degrees or warmer. They are not usually found near moisture. If found in kitchens they will be in a clock, refrigerator motor or similar area. They are often found around dresser drawers.

Van Waters & Rogers Inc.
subsidiary of Univar

German Cockroach

Color - Tan or light brown with two dark stripes behind its head.

Description - The female carries approximately 36-48 eggs per capsule. Capsules are carried until a few hours before hatching. They will eat almost any food consumed by man. This cockroach is most commonly found where food is handled or stored. Temperature preference is similar to man. They are more active than other cockroaches, and do so primarily at night. They are nocturnal or night creatures. Scavenger type feeding - will eat starches, sweets, meats, vegetation and grease. German cockroaches are occasionally known to be vectors of disease, especially intestinal infections. They are capable of carrying disease germs on the exterior of their bodies.

Life Cycle - One to two years or more depending upon climate conditions.

Home (Nesting Site) - Prefers warm and moist conditions. They are gregarious and live in groups. Frequently groom their bodies and prefer to feed at night and live in the dark. They do not crawl into a hole, but back into a hole, leaving their antennas sticking out to detect if there are insecticides in the air or any other kind of danger.

Oriental Cockroach

Color - Dark black or dark reddish brown.

Description - Female almost wingless, however, the male has wings a little shorter than its body's length. They are found more in the north than in the south. Most commonly found in dark damp basements or crawl spaces. Will ascend to rooms above. Eats or disfigures food, bookbinding and fabrics. Also, gives off a very bad odor. They pollute food and are possible carriers of diseases. Favorite foods are filth, garbage and decayed organic matter.

Life Cycle - One year.

Home (Nesting Site) - Often live in debris, such as leaves and garbage. Also, can be found outdoors in sod and vegetation around foundations and sewers.

Beetles

Black Carpet Beetle

Color - Dull black with brownish antennae.

Description - Oval in shape. Feeds on waste, grains, flour, powdered milk, candy, wool products and furs. Basically the same as other carpet beetles.

Life Cycle - One to three years or more.

Home (Nesting Site) - They have the same behavior as furniture and carpet beetles.

Cadelle Beetle

Color - Shiny black.

Description - Oval in shape. Feeds on waste, grains, flour, powdered milk, candy, wool products and furs. Basically the same as pantry moths and other grain pests.

Life Cycle - About one year.

Home (Nesting Site) - The same as other grain pests.

Cigarette Beetle

Color - Reddish brown to medium brown.

Description - Round beetle and breeds on tobacco. Also loves other dried plant products such as spices, red pepper, cayenne pepper, ginger, paprika, drugs, grains, dried raisins and cereal products. They can penetrate paper packaged materials.

Life Cycle - Up to one year.

Home (Nesting Site) - In cigars or other tobacco products.

Confused Flour Beetle

Color - Reddish-brown.

Description - Confused flour beetles cannot fly.
Prefers temperatures of 80 degrees. Within this
temperature their eggs can hatch within six days.
When temperatures are at 80 degrees, they can hatch
up to 300 or more eggs. When temperatures are
below 40 degrees, the eggs will die.

Life Cycle - From two to three years or more,
depending on the conditions of the climate.

Home (Nesting Site) - They can develop in flour and
crumbs. They can also be found in cracks and
crevices, including corners of cabinets and in voids.

Furniture Carpet Beetle

Color - Blackish with yellowish and white scales.

Description - When adults are present, this is the first sign of infestation. They are oval in shape and are sometimes found flying around lights, shades and in window seals. Infestations can be found in groups in corners of carpets and furniture. The adult will not do damage in its larva stage. They damage wool, feathers, dried meat, leather, hair and fur. Also, they will damage rugs and stored clothing in closets.

Life Cycle - Five months to a year.

Home (Nesting Site) - In rugs, mats, furniture, trunks, dresser drawers, pilling up of lint in floor cracks, rugs, basements and vents. Adults can be found also in window seals.

Saw-Toothed Grain Beetle

Color - Dark red to dark brown.

Description - They cannot fly. They feed on almost all dried food products. Will not attack sound grain. They destroy barley, breakfast cereals, corn, cornmeal, corn starch, flour, macaroni, oats and crumbs in furniture.

Life Cycle - One to up to three years.

Home (Nesting Site) - Around grains, in furniture and in cracks and crevices.

Crickets

Camel Cricket

Color - Camel cricket has a light to dark brown color. The field cricket is a dark brown to black. The house cricket is a yellowish brown.

Description - They can damage things and are a nuisance to people. They occur during the warm months of the year. They breed in areas kept damp such as shrub beds, grass, piles of leaves or in mulch. They will live in bricks, blocks and lumber. They are not indoor pests. They are brought in by too much moisture outside caused by rain. Adults are attracted to light in large numbers in the summer times. They are nocturnal or night insects. They eat holes in fur, clothes, paper and damage food.

Life Cycle - Up to one year.

Home (Nesting Site) - Lives in damp, dark and slightly warm areas in a structure. They love basements and closets.

Pantry Moths

Almond Moth

Color - Brown to grayish, with pale reddish bands across the middle of wings.

Description - Sometimes can be found in dried fruits, nuts and pet food. Their favorites are chocolate candies, nuts and dried fruits.

Life Cycle - Up to six generations per year.

Home (Nesting Site) - Inside of nuts and can be found in cracks and crevices. Same as other grain pests.

Angoumois Grain Moth

Color - Yellowish, gray to brown.

Description - Lays eggs in stored grain. They infest whole kernels of grain, especially corn. Larva attacks inside of grain.

Life Cycle - Minimum of five weeks.

Home (Nesting Site) - Inside of grains. Same as other grain pests.

Granary Weevil

Color - Dark brown to nearly black.

Description - The covers on wings have ridges on them. The head has a snout. They are very small -- about ⅛ of an inch long. The larva, pupa and eggs feed and develop inside corn kernels and grain. They will feed on flour, but will not produce eggs in it. They destroy food storage products such as spaghetti, macaroni, beans, corn kernels and other grain products.

Life Cycle - One to three months and up to two years, depending on the conditions of the climate.

Home (Nesting Site) - Inside grain products. They will live in and attack spaghetti, and other grains that are large enough for the larva to get into. This includes oats, wheat, rye and barley, and sometimes beans and nuts.

Indian Meal Moth

Color - Reddish brown with some gray on wings.

Description - It is the most common food infesting moth. They are common in grocery stores and other food establishments. The larva spins a silk like web over food. It attacks stored grain products, such as dried fruits, powdered milk, chocolates and other candies. When wings are folded, they look like a tepee. They destroy candies, wheat, flour, nuts, dried fruits, raisins, figs and prunes.

Life Cycle - Minimum of six to eight weeks.

Home (Nesting Site) - In foods and sometimes hangs around ceilings in homes and on the wall.

Lesser Grain Borer

Color - Dark brown or black beetle.

Description - Adult is a strong flyer. The larva feeds on flour, very fine grain or whole grain. They destroy flour, corn and grains. However, they will not destroy cereal products.

Life Cycle - Two months to one year, depending on climate conditions.

Home (Nesting Site) - In grains. Same as other grain pests.

Meal Worm

Color - Yellow brownish, brownish black, with circle rings around the body.

Description - They are scavengers and prefer to feed on decaying grain or spoiled milled cereal. They feed on cake mixes, corn-on-the-cob, corn meal, starch and grains. Others include oatmeal, bread and crackers. They destroy grains such as mixed foods, corn starch, corn meal, oatmeal, dried soup mixes, potatoes, bran and flour. Meal worms are also used for fishing bait and pet food.

Life Cycle - A year or more depending upon conditions.

Home (Nesting Site) - In dark damp places, spoiled grain products and in the basement. They can even live under old damp carpeting.

Rice Weevil

Color - Same as granary weevil, but has four red to gold spots on wing cover.

Description - The covers on wings have ridges on them. The head has a snout. They are very small -- about ⅛ of an inch long. The larva, pupa and eggs feed and develop inside corn kernels and grain. They will feed on flour, but will not produce eggs in it. They destroy food storage products such as spaghetti, macaroni, beans, corn kernels and other grain products.

Life Cycle - One to three months and up to two years, depending on the conditions of the climate.

Home (Nesting Site) - Inside grain products. They will live in and attack spaghetti, and other grains that are large enough for the larva to get into. This includes oats, wheat, rye and barley, and sometimes beans and nuts.

Clothes Moths

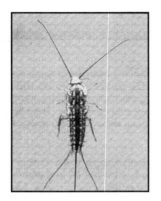

Firebrats

Color - Greyish, brownish with white to greyish and black patches.

Description - Firebrats prefers temperatures above 90 degrees. Often found around heating plants, ovens and other places that are extremely warm. They destroy book bindings, starched cloths, can labels, wallpaper, glue and paste.

Life Cycle - Their life cycle is from three to 24 months, depending upon the temperature and humidity.

Home (Nesting Site) - Any room where warmth and moisture prevails and a sufficient amount of food is found. Found in dark closets, behind baseboards and in walls. They like to stay around the baseboards.

Silverfish

Color - Silver-gray and scaly.

Description - The silverfish has two long antennas, three external tails, no wings and the body tapers from head to tail. They destroy book bindings, starched cloths, can labels, wallpaper, glue and paste. They like temperatures ranging from 72-80 degrees.

Life Cycle - Their life cycle is from three to 24 months, depending upon the temperature and humidity.

Home (Nesting Site) - Any room where warmth and moisture prevails and a sufficient amount of food is found. Found in dark closets, behind baseboards and in walls. They like to stay around the baseboards. Silverfish prefers warm and damp places such as basements and sometimes in attics.

Webbing Clothes Moth

Color - Shiny golden scales.

Description - Is not attracted to light, as do ordinary outdoor moths. Spins a silk like web on infested woolen material. They destroy wool, fabrics, brushes and fur. Similar to case making moth and carpet and furniture beetles.

Life Cycle - Two to three years on most depending on conditions.

Home (Nesting Site) - In dresser drawers and in closets as other fabric pests.

Flies

American Dog Tick

Color - Brownish with yellow markings on the shield. The adult female is bluish-gray in color.

Description - About ⅛ to 3/16 of an inch. Adults commonly climb up a piece of tall grass where they wait for a host animal. With three pairs of out stretched legs waving back and forth, they grab on and eat when the animal brushes by.

Life Cycle - Up to a year depending upon climate conditions.

Home (Nesting) Site - They are commonly found in recreational parks along the edges of grass, walkways and trails. They are drawn to the scent of animals.

Bed Bugs

Color - The bed bug is oval in shape and reddish-brown to reddish dark-brown to black in color. They are clear to very light brown in color when unfed.

Description - About 1/5 to ¼ of an inch. They have no wings, are blood sucking insects and have an offensive odor. Bed bugs hide in the day in beds. Specifically, they hid in mattresses, cracks in the floors, beds, windows and baseboards. They come out and feed on their host at night.

Life Cycle - About three months or more depending on climate conditions.

Home (Nesting) Site - Lay in cracks of furniture, behind baseboards, underneath loose edges of wallpaper, cracks around window frames, cracks in wooden floors and around beds. They can also be found in hotels, motels and in public areas.

Brown Dog Tick

Color - The adult male is a reddish-brown color. The adult female is grayish-blue to grayish-olive in color, with a reddish brown shield.

Description - The adult male is about ⅛ of an inch. The brown dog tick is an indoor pest and is active year round. Activity is at its peak in the fall. The brown dog tick is unable to survive in the northern states outdoors in the winter. They rarely bite man.

Life Cycle - Brown dog ticks live about four months or more.

Home (Nesting) Site - They tend to feed solely on dogs.

Flea

Color - Black or brownish black - very small.

Description - Flea eggs are dropped by the adult female, while feeding on pets. Adults are wingless and hard shelled. They feed on the blood of pets. Fleas can jump relatively long distances. Dog and cat fleas will even feed on humans. They breed throughout the year requiring 30-75 days to complete a life cycle. Adults emerge from the pupa, when an animal is near or when the carpet vibrates. They are sensitive to body heat. Eggs can hatch within two weeks. If pets are removed and there are severe infestations, the starved flea will attack man on the lower leg and ankles and on animals. This will cause itching, irritation and sometimes infections. Fleas can transmit bubonic plaque from man to man. They are vectors in the spreading of murine or epidemic typhus (fever causing disease) from rat to man.

Life Cycle - Adult fleas can live more than 18 months without a blood meal. Being susceptible to temperature changes, the life cycle can be prolonged. This varies among species and depends on the climate.

Home (Nesting) Site - When they are not feeding, they live in pet beds, sofas, carpeting, and areas cats and dogs use. Also, they live underneath beds and in closets. They can live in carpets up to one year.

Lice

Color - Colorless to grayish or light brown, but varies among species.

Description - A blood sucking or biting insect. They have no wings. They live on the bodies of animals and humans. True lice suck blood and spend their entire lives as parasites -- an organism living in or on another living organism and obtaining its nutriment partly or wholly from it.

The eggs which are called "nits" are usually attached to the hair of their host. There are three kinds of lice that can live on man -- the body, head and crab lice. Head and crab lice are much smaller than the body lice. Body lice are about ⅛ of an inch long. The common names for lice are the body lice, or "cootie" as the older generation called them, the head lice and the crab lice.

Life Cycle - Their life cycle will depend upon high humidities above 55° to high temperatures more than 75°F.

Home (Nesting) Site - Body lice lives in people's clothing. The head lice lives in the hair of people. Crab lice are found on the pubic hair of men and women.

Wood Tick

Color - Are somewhat clear in color, unfed, and are very small. After feeding, the larva color is like a bluish gray and dark gray in color. When the nymph is unfed, they are pale yellow brown, a gold like color and grayish after a meal.

Description - Ticks are closely related to spiders and mites. Ticks are blood suckers. The body is scale like. It has a head, thorax and an abdomen. All are one, with no segments. The larvae have six legs and the nymphs and adults have eight legs. Their jaws are like scissors and used for piercing the skin. They thrust their entire head into the wound once pierced. They have an anchor like toothed device below the jaws, that holds firmly to the flesh. Any attempt to pull the tick off will often break its head off and leave it buried into the wound. The female lays eggs, but in some species, the female carries their eggs until they are fully developed. Ticks lay their eggs in the ground, home or nest of their victim.

The most common tick is the wood tick. The wood tick attacks man, animals and rodents. They are carriers of diseases, namely the Spotted Fever and Tularemia (an infectious disease of rodents -- especially rabbits and are transmittable to man). The Rocky Mountain wood tick also carries the Rocky Mountain Spotted Fever.

Life Cycle - Adult ticks can survive at least one

year or more without a blood meal. They live about two years or more depending upon the species and climate conditions.

Home (Nesting) Site - Eggs are under barks, in grass, vegetation and in animal bedding areas. Eggs are in cracks and crevices of floors and walls, corners, on the edge of carpets, draperies, furniture and animal bedding.

Rats and Mice

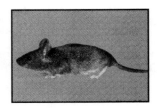

House Mouse

Color - Dusky gray fur with a small slender body.

Description - The young house mouse resembles the adult, but smaller. There are usually five to six young per litter. A house mouse can have eight litters a year. It can adapt itself to most weather conditions. Because of their size, they can get through openings, any opening the size of a dime, ¼ or larger and can be carried around in merchandise. The tail is equal to or a little longer than the body and head combined. The ears are large for the size of the animal. Weight is about ½ ounce to ¾ ounces. They destroy and pollute food products. Many are known to make holes in packages, gnaw on doors and structural objects. Mice transmit diseases, germs and human ectoparasites.

Life Cycle - Average of one year or less.

Home (Nesting) Site - Any room where food is stored, prepared or handled. Mice normally stay within a radius of 10-25 feet from their nests.

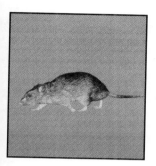

Norway Rat

Color - Coarse and reddish-brown.

Description - Norway rats have heavy set bodies with a blunt muzzle. The tail is shorter than the head and body combined. The ears are small and close set. The maximum weight is about one pound. The young rat looks like an adult, but smaller. They have a large head and feet. Their litter usually averages eight to 12 babies, with the female having four to seven litters a year. They destroy and pollute food products. Mice are known to make holes in packages, gnaw on doors and other structural objects. Bites are very painful. Rats sense with their body hair and whiskers. They use their whiskers to stay in contact with the surface of the wall. They are color blind, but have a keen sense of hearing. They can locate the source of a noise within six feet. Noises usually cause rodents to run.

Life Cycle - The length of life cycle averages one year.

Home (Nesting) Site - They live in the ground, by digging burrows in rooms, buildings where food is stored, prepared and handled. Also lives in basement, crawl spaces, dumps and garage areas. They harbor in the ground level of structures, in holes in the ground under buildings, porches, houses, storage areas and closed places.

Roof Rat

Color - Three color varieties - black to slate gray, tan above and grayish-white below, or tan above with yellow or white belly.

Description - The muzzle of the roof rat is slender and pointed. The tail is longer than the head and body combined. The ears are large and easily seen. Weight is about eight to 12 ounces. Roof rats can jump nearly two feet vertically, three feet with a running start, four feet horizontally, and eight feet from an elevation that is 15 feet above the finish point. Rats can reach upward about 18 inches.

They are excellent climbers and will often walk along telephone wires to enter homes and buildings. They may also use over hanging limbs. Roof rats often enter from above ground. They are not controlled by baits and traps on the ground level. The best way to catch them is to use traps up in attics.

Life Cycle - The length of their life cycle averages one year.

Home (Nesting) Site - Any room or area of building, crawl space, attic, particularly where food is stored, prepared or handled. They like to climb, often living in trees or palm trees. They nest above ground level, preferably in attics, between walls, enclosed spaces of cabinets, shelves and trees.

The roof rat is capable of living in uninhabited areas. They survive well on fruits and other foods found in forests and other outdoor areas. They are more likely to be found in rural areas, but are also found in cities.

Mosquitoes

House Fly

Color - Dull gray with stripes on its back.

Description - Flies cannot eat solid foods. They will regurgitate and then eat. This process breaks down the food. How they transmit diseases are by their mouth parts, legs, through their vomit, or through their feces. They are a nuisance to man. The diseases they transmit are typhoid fever, tuberulosis, diarrhea, dysentery -- a painful intestinal inflammation characterized by diarrhea with bloody and mucous feces -- cholera and salmonellosis (food poisoning). House flies have filthy habits. They pick up diseases and germs during their travels over garbage, sewage and other sources of filth. Then, they transmit these germs to human and animal food by their mouth especially, their vomit and feces.

Life Cycle - Under good conditions they can produce a new generation in seven to 10 days. However, under normal conditions it takes 10-18 days to complete the life cycle. Adult flies can live up to 60 days.

Home (Nesting Site) - During the day, flies will rest on floors, walls, ceilings, tables, plants, bushes, utensils and garbage areas which supply them food. Garbage areas are a good source for breeding. Night resting areas are near daytime feeding areas. They spend most of their time near their breeding areas, but will travel great distances in search of food.

Mosquitoes

Color - Varies among species.

Description - Mosquitoes have scales on their wings and body. The average size is about 1/5 of an inch long. Mosquitoes have a head, body (thorax) and abdomen. The head has a long tube like mouth called proboscis and a pair of antennas. Mosquitoes have long and narrow abdomens. They are persistent biters. Bites are very irritating and can transmit diseases.

There are about 2,000 or more mosquitoes known to man. They are divided into about 100 or more groups or kinds. Some of the ones that are known in North America are the culex, the aedes and the anopheles. The culex and the aedes are relatively harmless in the United States. There are species of the culex and the aedes that carry diseases, but are usually confined to tropical and subtropic areas. There are two of the species of anopheles that transmit malaria found in the United States. Mosquitoes are true flies, having only one pair of wings.

Life Cycle - Varies among species and climate conditions, about one week or more.

Home (Nesting Site) - They breed in water, streams, rain barrels, hollow stumps of trees, cesspools, drainage ditches, puddles, flower vases, old toilets, old pools, bird baths, fountains and in other areas where water stands undisturbed for a long period of time. The adult then emerges from the water and flies away to look for food or blood from humans and animals. They can lay up to 300 or more eggs that float on the surface of water. The eggs hatch as larva in a few days.

The larvae are called wrigglers, because of the way they move in water. They frequently seek food in the water and come up to the surface of the water to breathe. In a few days, the larva is then developed into the pupae stage. In the pupae stage they do not eat. Some mosquitoes are active in the day and rest at night.

Many species of the culex and the anopheles are active at night and rest during the day. The adults stay hidden during the day in outhouses, shrubbery, cars, trucks, homes, and basements. They bite in the late afternoon and early morning.

Slugs and Snails

Slugs

Color - Grayish brown to a dark grayish which varies among species.

Description - Slugs have small and flat shells under their skin, but most of them don't, unlike snails which have a shell on the outside. They have two antennas with eyes on the outer end of the longer pair. The other pair is sensitive to touch. Because of its very large appetite for eating plants and other foliage, slugs become a pest. They can grow up to four inches or more depending on the species. Long in length, they are slow moving. The mouth is in the center of the head. They hide in damp and moist areas in dark places in the day. Slugs come out at night for food. Depending on the species, some will eat insects and worms. Most, however, eat plants and destroy gardens which makes them a pest. In the winter, they hibernate under ground. They can be a nuisance when they are on sidewalks and steps.

Life Cycle - About one year depending on species and climate conditions.

Home (Nesting Site) - Hiding places may include underneath damp and wet boards, logs, basements, rocks, plants and leaves.

Snails

Color - Grayish brown to dark grayish.

Description - Snails have antennas with eyes on the outer end of the longer pair. The other pair is sensitive to touch. Because of its very large appetite for eating plants and other foliage, snails become a pest. They can grow up to four inches or more depending on the species. Long in length, they are slow moving. The mouth is in the center of the head. They hide in damp and moist areas in dark places in the day. Snails come out at night for food. Depending on the species, some will eat insects and worms. Most, however, eat plants and destroy gardens which makes them a pest. In the winter, they hibernate under ground. They can be a nuisance when they are on sidewalks and steps.

Life Cycle - About one to two year or more depending on species and climate conditions.

Home (Nesting Site) - Hiding places may include underneath damp and wet boards, logs, basements, rocks, plants and leaves.

Ants

Argentine Ant

Color - Brown.

Description - When crushed, the Argentine ant has a musty like odor. Trails are clear and can have many queens in one nest. Do not always have mating flights. They feed on sweets, fats and oils.

Life Cycle - From one to 15 years, depending on ant species and climate conditions.

Home (Nesting) Site - Found in the southern parts of the United States. They rest outdoors under logs, concrete slabs and other debris. In the winter they will move indoors.

Carpenter Ant

Color - Red to black.

Description - Larger than most ants. Found in hollow trees and in wood debris. Sometimes they nest underneath bark or wood chips. They do not eat wood, they hollow through the wood to form a nest. They eat many foods such as sweets, meats, grease and dead or alive insects. In the spring or early summer, when the weather conditions are favorable, about 200 or more swarmers emerge for mating flights. This occurs for several days. Fertile queens then seek some suitable locations to begin new colonies. They prefer to attack moist or partially rotted wood than sound wood since it is easier to excavate.

Life Cycle - Male dies after mating. The colony is mature when reproductives are formed. This usually occurs within three to six years.

Home (Nesting Site) - They will nest in natural voids, where very little excavation is required. They will live in dead tree limbs, rotted logs, wood piles, wood porches, fences, window frames, shingles and siding. Carpenter ants also live in wooden barks around homes.

Carpenter Bee

Color - Black and blue-like purple with a metallic sheen.

Description - Carpenter bees closely resemble bumble bees in size and appearance. The difference can be distinguished by the lack of hair on the upper body which is called the abdomen. Carpenter bees are more bare and flattened -- from ¼ inch of an inch to one inch in length.

The female makes the nest and the male helps to protect it though they are unable to sting. The female stores a supply of food at the end of the tunnel and lays an egg on the food. She then seals off the cell with tiny pieces of wood. The wood is stuck together with a glue-like saliva. This process is repeated until the tunnel is filled with cells. Each contains a good supply for the larva. When the bees emerge from the nest, they will have cleaned the nest out. One of the young will return to the nest to start the process all over again.

Life Cycle - One year. The female does not survive the winter.

Home (Nesting Site) - In the heart of wood.

Harvester Ant

Color - Red to dark brown.

Description - About 3/16 of an inch to ⅜ of an inch long. You can see its stinger on the end of its abdomen. Primary food is seed. They gather and store certain grass seeds or grains cultivated in an underground chamber. On sunny days they bring the seed out to dry.

Life Cycle - From one to 15 years depending on ant species and climate conditions.

Home (Nesting Site) - Under ground, concrete slabs and under wood piles.

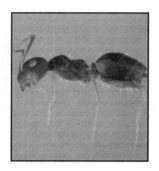

Honey Ant

Color - Reddish brown to shiny black.

Description - The adult workers are about 1/16 of an inch to ⅛ of an inch long. Antennas do not have clubs at the end. The body is smooth. They eat foods like live and dead insects, juices from fruits and certain plant parts or seeds.

Life Cycle - From one to 15 years depending on ant species and climate conditions.

Home (Nesting Site) - They hang from the ceiling of nest chambers and dispenses food to the other ants during the dry season.

Little Black Ant

Color - Black.

Description - About 1/10 of an inch long. Body is shiny black and slender. It can have several queens in each colony.

Life Cycle - From one to 15 years depending on ant species and climate conditions.

Home (Nesting Site) - They are found in houses and lawns.

Pharoah Ant

Color - Yellowish to light reddish clear color.

Description - About 1/10 of an inch. It has three segments on its antenna at the end and two segments in the pedicel before the abdomen. Can have many queens in each nest. Pharaoh ant colonies can have up to 300,000 or more individuals and many queens. Reproductives do not swarm, as mating occurs in the nest. They establish new colonies when the nest begins to become over crowded. Then, a group of queens and workers split off from the mother colony. When a residual insecticide is used, the colony is then forced to split up. If the chemical fails to destroy or come in contact with the nest, they will then stay closely together until the chemical barrier breaks down. Once the barrier breaks down, they will split up into several colonies which then would be hard to control. Pharaoh ants are very intelligent ants.

Life Cycle - From one to 15 years depending on ant species and climate conditions.

Home (Nesting Site) - Pharaoh ants can be found outside in warmer areas. Normally they are found indoors in buildings, walls, wall voids, ceilings, between floors, in foundations, hotels, large apartments, homes, groceries and in other places where food is commercially prepared or handled. These ants are often in hospitals and are very difficult to treat in these areas. Preferred nest locations include warm areas near fireplaces, hot water pipes, furnaces and heat ducts.

The colony water needs are high. Workers are often seen around sinks in kitchens and bathrooms getting water. Pharaoh ants prefer foods containing fats and oils and are classified as grease feeders. They will, however, utilize almost anything edible as a food source. Trails may extend considerable distances from the nest, and are concealed as much as possible. Trails usually follow electrical conduits and water pipes (plumbing) through wall voids. Lives in walls of buildings.

Southern Fire Ant

Color - Yellowish-red.

Description - From 1/16 of an inch to ¼ of an inch, they are long and shiny. Smaller ones are a little darker. They have prominent eyes that are visible, possess a hard body and stings like a bee.

Life Cycle - From one to 15 years depending on ant species and climate conditions.

Home (Nesting Site) - They are found in the southern parts of the United States. They usually have mounds outdoors in sunny areas of the lawn or woods.

Thief Ant

Color - Light yellow to a clear like bronze.

Description - One-fifth of an inch to 1/10 of an inch. There are two segments on their antennas. They have very small eyes. Thief ants can have many queens in a nest. Their primary food is meat.

Life Cycle - From one to 15 years depending on ant species and climate conditions.

Home (Nesting Site) - They will nest near other ant colonies to steal food and larvae to feed their own colony. Outside they nest under rocks, wood piles and other debris. Inside, they can be found in wall voids behind baseboards, kitchen cabinets and plumbing areas.

Beer

Bumble Bee

Color - Usually black with yellow or light orange markings and are covered with very thick short hairs.

Description - Large, strong and rough insects about ½ inch to 1-¼ inch long. They make their nests in clumps or piles of dry grass and in abandoned burrows or holes in the ground. They use the same system for the queen like the honey bees.

Bumble bees feed on pollen and nectar from plants, namely flowers. Bumble bee colonies do not survive the winter. The queen hibernates or sleeps in her burrow, but the rest of the family dies. The queen selects a new site for a nest, usually in the spring. The queen lays eggs in the cell and soon the larvae hatch. The larvae develop into the first workers. The workers then take over the work of gathering pollen and nectar. The queen then makes it her sole duty of laying eggs. Late in the season, a few larvae develops into new queens. Drones are produced to make the queen and carry on the yearly cycle. Bumble bees do not produce honey for the consumption of humans, but pollinates plants which are of great importance. They are the only bees with tongues long enough to reach the nectar of red clover blossoms.

Fertile queens leave the colony in the fall to locate

a protected area in the winter. Queens will establish a new nest in the spring and raise a small brood or family of workers entirely on her own. Once workers have emerged, they tend to the nest and the queen limits her activities to egg laying.

Life Cycle - About one year.

Home (Nesting) Site - In clumps of dry grass and in burrows in the ground.

Honey Bee

Color - Pale yellow to amber brown.

Description - Honey bees are raised by man. We use its honey as a sweetener in syrups, candies and baking. The honey bee produces a wax called bees wax which is used to make candies, floor waxes, polishes, plasters and ointments. Honey bees pollinate about 80 percent of crops raised by man. Many farmers who grow fruit rent honey bee hives to improve pollination during the flowering seasons. Honey bee colonies last through the winter. They live on the honey that they have stored during the winter. They keep warm by surrounding the queen and beating their wings.

The honey bee is not a native of North America. They were imported from Europe. The most widely raised honey bees in America is the Italian honey bee. The Italian honey bee is a more gentle honey bee. There are also other varieties of bees, which includes the Laucasin from Russia, the Carnelian from Australia, the German and the African bee which are popularly called the "killer bee."

Life Cycle - Worker bees live six weeks to three months or more. Depending upon climate conditions, queens live about three to five years or more.

Home (Nesting) Site - In hollow trees, logs, caves or in crack and crevice openings under a ledge of a house. They also live in attics and wall voids.

Queen Bee and Drone Bee

Color - See bees.

Description - The queen is a little larger than the other bees. Her stinger is smooth and curved, which is used only in battles with rival queens. She can live up to three to five years, during which her whole time is devoted to producing eggs. She will lay one egg per minute in the summer.

The drones are a little smaller than queens, but larger than workers. They have no stingers and do no work. When mating season is over, they are usually driven out from the nest or killed by the worker bees.

Life Cycle - Three to five years.

Home (Nesting) Site - Same as the other bees listed in this reference book.

Paper Nest Wasp

Color - All species are banded with yellow on a black or brown background color.

Description - They have stingers, build paper nests and form yearly colonies. They have short, heavy antennas and most have a thick waist. Paper nest wasps are social insects with both fertile and infertile females and males. The fertile males and females live longer in winter hidden in cracks of trees, attics or other secluded places. In spring, they mate and build a nest in which workers are raised. The queen begins the nest when she emerges from hibernation in the spring. The queen makes a few cells, glues an egg to the top and feed the young when they emerge in about eight to nine days. In about 12-14 days, they become pupae. In another nine to 11 days they become adult workers. The queen confines herself to laying eggs.

Life Cycle - One season with new fertile males and females.

Home (Nesting) Site - After mating with the males, the females go into hibernation. Usually this is inside a building of some sort for the winter. Those that survive become the new queens of the new colonies in the following spring. The cold weather kills the rest of the colonies.

Solitary Wasp

Color - Blackish-purple color, but varies among species.

Description - They live alone, and most of them never see their parents or the young they produce. They do not live in colonies and there are no queens, drones or workers. Nests are made with damp soil. The nest cells are supplied with stunned insects or spiders before egg laying and then sealed afterwards. When the eggs hatch, the larva feeds on the contents in the cells and later emerges from the cell as an adult. The adults like other wasps have the same narrow threadlike waist.

Life Cycle - About one year.

Home (Nesting) Site - All the females are fertile and they construct the nests and provide for the grubs. The eggs are laid in tunnels or holes in the ground. Some males survive the summer, others die a few weeks after mating. With its stingers, the wasp injects poison into certain nerve centers of its victim. The poison keeps its victim quiet without killing them, insuring the larva a safe supply of fresh food.

Wasps

Color - Varies. See hornets, yellow jackets and paper wasps.

Description - Wasps are stinging insects related to bees and ants. There are thousands of different kinds of wasps and new ones are discovered year round. The males do not have stingers.

Although wasps do not commonly collect nectar or pollen, they do drink nectar when going from plant to plant to obtain nectar. They spread pollen. Very few wasps make honey. Many species injure ripe fruit and some wasps are pests. Several species do damage by killing bees and other insects such as spiders. Wasps really do more good than harm by getting rid of pests and pollinating plants.

Life Cycle - About one year.

Home (Nesting) Site - They live in nests made of paper like material. This material is made of bits and pieces of weathered wood (wet or very moist) or bark mixed with saliva.

Worker Bee

Color - Varies among species.

Description - The workers collect bees wax secreted by their bodies. From it they build small six sided cells for storing honey and pollen. It also serves as heated cells for eggs, larva and pupae. These cells are joined to form a thick wax shaft called a honeycomb. Heated cells for male eggs are larger than worker eggs.

An adult worker grows from a newly laid egg in about 20 days. The first duties of young workers are to feed larvae still developing in other cells. After that, the worker guards the entrance of the nest and fans their wings to circulate air.

A group may make 20,000 to 40,000 separate trips in collecting enough nectar to make one pound of honey. A single colony may produce up to 500 pounds of honey. Workers communicate by dancing. By dancing in various patterns, a bee can inform the other workers of the direction and distance from the hive of a food source and suggests its richness.

The worker bee develops a new queen when a hive becomes over crowded. They do this by providing a female larva with a diet of royal jelly. A mature queen will grow from a fertilized egg in about 16 days. If more than one queen matures at the same time, they will battle among themselves with their stingers until only one remains alive. Then the new queen begins her mating flight. She goes to flight in the air, flying sharply upward, attracting drones from her own hives and hives close by. She outdistances all, but the strongest drone becomes her mate while in flight. The drone dies immediately and she returns to her nest to begin laying eggs. The worker bees then surround her, watch over and feed her.

Life Cycle - One to six months or more depending upon the species and climate conditions.

Home (Nesting) Site - The old queen leaves with about half the workers. She leaves the newest to start a new colony when a new queen develops. The bees then swarm around the queen making the queen its main attraction. When the swarm enters the new nest, food is immediately gathered and a new honeycomb is built. The workers then secrete saliva or wax from their glands on the abdomen, mold it with their mouths to form combs. The queen then deposits her eggs in these combs.

Yellow Jacket and Hornet

Color - Hornets' bodies are black and yellow or black and white. Yellow jackets are ½ inch long in length with black and yellow markings.

Description - Yellow jackets and hornets are wasps with strong rough bodies that build large nests of paper construction. This paper material is made from bits and pieces of not very moist weathered wood or bark mixed with saliva and worked consistently with the jaws. They are attracted to all types of fruit. Hornets may vary from about ¾ of an inch to 1-¼ inch. Large colonies can have up to several thousand workers.

Life Cycle - About one year depending on species and climate conditions.

Home (Nesting) Site - They will nest in the ground, hollow trees, wall voids or attics. Their layered nest is covered completely and is commonly attached to a tree limb or on the side of a home, building and sheds. Nest sizes vary but can reach sizes larger than a baseball and contain several hundred workers. Adult workers are common pests wherever sweets are found at family picnics, homes and amusement parks.

Termites

Zootermopsis
angusticollis

Damp-Wood Termite

Color - White and light yellow soldiers having a brown head.

Description - The Damp-wood termite lives only in damp moist wood. These termites cause damages on the Pacific coast. It lives in galleries (in high areas) and chambers (private areas such as attics where dampness will occur above ground level). They drill-eat living woody plants or damp and rotting damp wood.

All termite classes contain both sexes. The king lives as long as the queen. There are three basic classes in most termite colonies -- reproductive, workers and soldiers. In some species, there is only one class of reproductive termites. It consists of a queen and a king, fully developed or as a perfect male or female. They have fully developed eyes, moderately hard bodies and their wings are fully developed. The queen termite is the largest.

Life Cycle - Queens can live up to 50 years.

Home (Nesting) Site - This insect lives only in damp and moist wood.

Dry-wood Termite

Color - They are usually dark in color or tan.

Description - The Dry-wood termite needs little or no moisture at all to survive. They do damage in the southwest and western states and live in abundance in tropical warm regions. These regions include Africa, Australia and the Amazon region. They can build very huge mounds sometimes 200 feet in height or even higher.

All termite classes contain both sexes. The king lives as long as the queen. There are three basic classes in most termite colonies -- reproductive, workers and soldiers. In some species, there is only one class of reproductive termites. It consists of a queen and a king, fully developed as a perfect male or female. They have fully developed eyes, moderately hard bodies and their wings are fully developed. The queen termite is the largest.

Life Cycle - Queens can live up to 50 years.

Home (Nesting) Site - They infest dry and dead wood and wood objects that are made by man. They also build shelter tubes by using digested wood matter. They do not use dirt or soil.

Subterranean Termite

Color - Milky white workers. Soldiers are white to light yellow with a dark brown head.

Description - The Subterranean termite is the smallest of termites and the most destructive. They will build mud tubes of soil mixed with waste matter as runways from their nest to the wood or its food source. They use these tubes for shelter which protects them from being detected and from sunlight and dry air.

All termite classes contain both sexes. The king lives as long as the queen. There are three basic classes in most termite colonies -- reproductive, workers and soldiers. In some species, there is only one class of reproductive termites. It consists of a queen and a king, fully developed as a perfect male or female. They have fully developed eyes, moderately hard bodies and their wings are fully developed. The queen termite is the largest.

Life Cycle - Queens can live up to 50 years.

Home (Nesting) Site - Their nests are located under ground and can extend their burrows for considerable distances into wooden structures. They will invade plants and wood structures that are in contact with the ground.

Banded Garden Spider

Color - The color is pale yellow with black lines on the abdomen.

Description - Is normally found in gardens, around homes and in tall grass. The legs are spotted.

Life Cycle - About one year or more.

Home (Nesting) Site - In gardens and around houses.

Black Widow Spider

Color - The adult female has a jet black shiny body with a red hour glass shape underneath the abdomen. The adult male is smaller than the female, usually has light marks on his abdomen and may have yellow marks on the top part of his abdomen. The spiderlings are reddish brown in color with light and dark marks on the abdomen and legs.

Description - Black widows live in cool, dark and damp places and are extremely shy. Body size is about ½ inch long and legs are about ½ inch long. Females will usually eat most of the young when they hatch. The female will deposit 200 or more eggs. The eggs are guarded by the female. She is the most seriously venomous spider in the United States. Her venom is a neurotoxin, that acts on the victims' nervous system.

Life Cycle - About one year or more.

Home (Nesting) Site - Under stones, gas pits, water pits, logs in dark caves and in dark areas of homes.

Brown Recluse Spider

Color - The adult is brown to tan or fawn -- light yellowish brown -- to chocolate brown.

Description - About ½ inch long. It can be identified by the distinctive violin shape mark on the cephalothorax (head joined to the thorax-chest). They do not bite unless molested, threatened, hurt or frightened. They can inflict a dangerous bite causing formation of a gangrenous lesion, although the initial pain associated with the bite is not intense. Within eight to twelve hours, the pain becomes very intense. Over a period of a few days, a large ulcerous sore will form. The bite itself is not fatal. However, a secondary infection will enter in. The ulcerous sore heals very slowly and leaves a very ugly, disfiguring scar. They can be found mostly in the eastern United States, but have been found in other states.

Life Cycle - About one year or more.

Home (Nesting) Site - Shy, lives in dark corners such as bedrooms, closets, in cracks and crevices, in barns, garages, crawl spaces, outdoor toilets and basements. Normally found near the ground or in corners.

Grass Spider

Color - Yellow to brown, with a pair of dark lines on top of the body. Dark marginal lines on its sides.

Description - They can be found around buildings. Size is about 1/16 of an inch.

Life Cycle - About one year or more.

Home (Nesting) Site - In wooded areas and around buildings.

House Spider

Color - Bodies are dull colored gray with dark spots and lines.

Description - Normal length of the house spider is 3/16 to 5/16. Female spiders lay approximately 250 eggs at one time, but may lay 3,760 eggs in her lifetime. Randomly selects its web site, to see whether it can capture its prey. If this web is unsuccessful in catching food for the house spider, it abandons the web. This pest is considered a nuisance because of its numerous spider webs.

Life Cycle - About one year. Survival is low in modern homes with low humidity.

Home (Nesting) Site - In upper corners, under furniture, in closets, angles of window frames, basements, garages and crawl spaces.

Jumping Spider

Color - The jumping spider has a color likeness like the black widow spider. Some have white, yellow or orange spots.

Description - They will bite, but the small amount of venom causes only mild irritation. They creep up and pounce or leap on their prey. They have short legs and can jump more that 30 or more times the length of their bodies. They have great skill. They possess the best eyesight of all spiders.

Life Cycle - About one year or more.

Home (Nesting) Site - Around homes, on bushes and exterior of buildings.

Orange Garden Spider

Color - Black with yellow or orange markings. The cephalothorax is gray above and pale to yellow beneath.

Description - The body is one inch long.

Life Cycle - About one year or more.

Home (Nesting) Site - In gardens, on bushes and around homes.

Tarantula Spider

Color - They are generally brownish in color, but the males are usually darker to the point of being black.

Description - Tarantulas look very fierce because of their very large hairy bodes. They are generally harmless to man. The bite is no more harmful than a bee sting. They received their name from a large wolf spider in Italy. They can grow up to three inches in body length and have a leg span of 10 inches. If handled improperly, they may inflict a painful bite but is not fatal to man.

Life Cycle - Some tarantulas live more than 15 years.

Home (Nesting) Site - Many tarantulas dig burrows for nests. They are found in warm climates like the southern and western states and in Central and South America.

Trap-Door Tarantula Spider

Color - Dark brown to bluish-black.

Description - This spider gets its name because it covers the entrance to its burrow with a lid which is made of silk threads, twigs, dirt or cork. The spider waits under the lid until an insect or other prey passes by. This spider can grow up to an inch long. They build a tube-like nest about four to six inches into the ground and line it with silk.

Life Cycle - Can live up to 15 years or more.

Home (Nesting) Site - They nest in a tube-like nest in the ground with silk webs over it. The nest has a trap door that is used for the spider to wait in hiding for its prey.

Wolf Spider

Color - Brown to dark brown to black.

Description - They are very excellent hunters. Many kinds have hairy bodies and are very swift runners in search of food. They are large spiders. The female wolf spider carries her young on her back (up to 80 or more) until they can care for themselves.

Life Cycle - About one year or more.

Home (Nesting) Site - In the day, some hide under stones or wood or in grass. Others are active in the sunlight.

Beneficial Pests

Ant Lion

Color - White to whitish grey with ridges like spun on back.

Description - The ant lion is an insect whose larva preys solely on ants. The body of an adult ant lion can reach from two inches to three inches long or more. The ant lion has two pairs of legs and slender wings. The antennas are short with knobs on the end. The larvae of the ant lion body are flat, tapering downward to the back. The larva is also called "doodle bugs."

The larva of the ant lion digs a pit like a funnel in the sand or soft loose soil and hides at the bottom of the pit. When an ant or other small insects fall in, the ant lion captures it and eats it. They have very strong jaws.

Life Cycle - Seasonal pests and may be found in other warm climates.

Home (Nesting Site) - Ant lions nest in the sand or soft loose soil. They also live in dry like sand under trailers, crawl spaces and other dry sandy areas.

Dragon Fly

372 Climber Dragonfly, 1¾–1⅝", p. 369

Color - Normally brightly colored with yellow, green, blue or red spots or stripes that stands out.

Description - Dragon flies can be called by two names, "dragon fly" or "darning needle." They are named for large beneficial insects that feed on harmful insects, like gnats and mosquitoes. There are 500 or more known in the United States. They are found in the region of swamps and streams of moderate climates and tropical zones. Dragon flies have large heads, bulging eyes and very strong jaws. They also have a long slender abdomen with three legs.

Damsel flies are smaller than true dragon flies and are very fragile. They seize insects by hovering (flying in a still motion) reeds and grasses near ponds. True dragon flies are larger and stronger. They are very fierce and pursue their prey in the air. They use all of their legs to seize their prey and crush them with their strong jaws. True dragon flies drop their eggs on top of the water.

Life Cycle - Depends on the species. This is a seasonal pest who lives in hibernation and in the larvae stage. Some are found in temperate areas.

Home (Nesting Site) - Damsel flies lay eggs on soft tissues of plants, just beneath the water surface. Both damsel and dragon flies live and play in the water, giving them the name, water nymphs.

Ground Beetle

Color - Dark brown to black.

Description - The ground beetle is a long, flattened beetle. They are fast runners and destroy harmful insects, making them helpful to mankind.

Life Cycle - Depends on the species, but they are seasonal pests.

Home (Nesting Site) - They hide during the day underneath rocks, wood and foundations. They can also be found around homes and wooded areas.

Lace Wing

Color - Light-pale green, golden eyed insect with very large translucent (clear image) lacy wings.

Description - Another name for the lace wing is "goldeneye." There are about 50 or more lace wings that are native to the United States. Size is about 4/5 of an inch long. Many lace wings are active at night, and are seen around lights and on lighted windows. They lay their eggs on plants. The young come out from the larva in about four to five days and begin eating soft insects and their eggs.

Life Cycle - Depends on the species, but they are seasonal pests.

Home (Nesting Site) - They nest on plants, shrubs, around homes and wooded areas.

Lady Bug

Color - Colors vary from red, yellow or orange and are spotted with black, white or yellow spots.

Description - The lady bug is often called the "ladybird." It was given this name due to its usefulness to farmers and gardeners. They are beneficial to man, both larva and adult. They eat large amounts of aphids, scale insects like mites, mealy bugs and other soft bodied insects that destroy plants. They can eat up to two dozen insects each day.

There are over 3,000 species of lady bugs. Their bodies are round dome-shaped and ⅜ of an inch long. Many species hibernate in the winter and some are all season. They are used to help farmers on farms and orchards. Some are used in other countries to help destroy non beneficial insects from destroying citrus plants. This includes the cotton-cushioned scale and the citrus mealy bug.

Life Cycle - Depends on the species, but they are seasonal pests.

Home (Nesting Site) - They nest on plants, shrubs, around homes and in wooded areas.

Praying Mantis

Color - Varies from green to greenish brown to brown coloration.

Description - Larvae and adults will feast on cutworms, beetles, flies, aphids and insect eggs. Each consumes many times its own weight. They relate it to the grasshopper and cricket. How the insect received its name comes from the way the insect's forelegs resemble praying hands. Their forelegs are very strong and are used for grasping their prey -- giving thanks. There are probably more than 2,000 praying mantises and many other kinds that we have not discovered. They are originally native to tropical and warm climates.

There are many species and the size can vary from about less that one inch to three inches long or more. Praying mantis has long hind legs and an oval shape leaf like wings. They lay up to 400 or more eggs on twigs in a frothy foam-like substance. Later it becomes dry and paper like. The young which have no wings hatch in the spring. When the young hatches, they feed on soft bodied insects like caterpillars, aphids and other insects that have soft bodies.

Life Cycle - Depends on the species, but they are seasonal pests.

Home (Nesting Site) - They nest mainly in tropical climates, around homes and wooded areas.

Treating Common Household Pests

RECOMMENDATION OF PESTICIDES TO BE USED

Whether from the store or from a pest supply warehouse, there are many products on the markets. Ask yourself the following questions. Which one do I to choose from? Which one really works? Which one is the best? The most important question to ask is the cost of the products. The answer to the question is, it is not the name that makes the pesticide good or work, it is the ingredient. Use the following as a guide.

◆ If a pesticide has *Pyrethrum* in it, you know then that this particular product has a "flushing agent" in it. It flushes bugs out of their hiding place or out of their nest (home).

◆ *Pyrethrum* is a naturally occurring pesticide derived from a source from plants. These are called "botanicals."

◆ *Boric Acid* is another good pesticide which works very well. Most boric acid seen in stores is good. However, professional boric acid can be purchased from a pesticide warehouse. Boric acid usually has a residual life span. When used in cracks and crevice areas, it lasts from six to eight months depending on the area being treated. Moisture can cut down the effect of the boric acid.

Boric acid can only be applied in cracks and crevice areas such as behind baseboards, inside walls, around kitchen pipes (plumbing) and in bathrooms. Boric acid should not be placed in open areas where children and pets can reach them or come in contact with. It is a naturally occurring insecticide that is derived from a source of minerals. It is also a stomach poison which kills insects slowly by interfering with the conversion transformation of energy within the insect's cells.

♦ *Fogging* - When using foggers from the store or professional foggers from a pest supply warehouse, always read the label before using. Make sure there are no pets present. Make sure there are no windows opened. Check and make sure all ventilation systems, gas stoves and fireplaces are turned off. This will help keep the fogging aerosol inside and eliminate mishaps. Make sure there are no birds present. For pets, make sure reptiles are kept out of the house or in a closed area. Further, make sure fish tanks are unplugged or the motor is turned off -- cover the fish tank with plastic or a piece of fabric.

♦ *Residual Pesticides* - Liquid pesticides are called "residual pesticides." It comes in contact with the surface and remains on the surface for a period of 30-45 days. This depends on where it has been applied or temperature conditions.

- ***Insect Growth Regulators*** - There are other insecticides in a liquid form which do not kill the insects. It, in turn, causes sterilization in adult insects. These insecticides are called "insect growth regulators."

- ***Restricted Pesticides and Nonrestricted Pesticides*** - You can purchase some of these pesticides from a pest supply warehouse. Some pesticides are for licensed professionals only. These pesticides are called "restricted."

    "Non-restricted" pesticides can be purchased by almost any one, but must be used as the label recommends.

Here are some pesticides that can be purchased by a non-licensed person. Each state is different, so be sure to check with your Department of Agriculture. *"Never apply any pesticide without reading the label."*

- Tempo - Wettable Powder
- Knox - Out (Diazanon) - Micro-Encapsulated
- Dursban - Residual Liquid
- Baits - Max Force Roach and Ant Bait by Raid
- Dursban and Diazanon - Granuals

Safety Tip - Always make sure all items such as toys, pet food and water, and dishes are picked up before spraying or treating the carpet or lawn.

"The Most Important Note" - Always wash your hands after handling or using pesticides, and before eating, drinking and using the bathroom. You need to protect your mouth and

private areas. These are very sensitive areas that are the first to come in contact with your hands. Therefore, this will decrease the risk of pesticide exposure.

Cockroaches

American, Brown-banded, German and Oriental Cockroaches

Always do a good inspection. This is the key to eliminating cockroaches and locating their nests.

- Look for evidence of live roaches.
- Look for cast skins (dead skins).
- Look for cockroach droppings - fecal matter.
- Look for regurgitation spots.
- Look for empty egg cases - capsules.

Be thorough - all harborage areas must be found and treated. These areas are usually found in plumbing areas (i.e., kitchens, bathrooms and under sinks). They can also be found in bath trap panels, behind showers and on the side of walls.

Plumbing areas provide deep voids and moisture for comfortable breeding for cockroaches. German cockroaches can also be found behind refrigerators, cabinets, stoves and appliances -- anywhere where the humidity is high. Cockroaches spend 75% of their time in harborage areas and 25% of their time looking for food.

In order for cockroaches to be successfully controlled, applications must be done to cracks and crevices. This will help maintain immediate control. Crack and crevice applications include treating behind kitchen and bathroom cabinets, underneath sinks, behind stoves, behind baseboards and dish washer machines.

Further applications should be applied to inside wall voids, which are called harborage areas. This includes dusting with an insecticidal dust inside of plumbing areas and will help eliminate 98% of the cockroaches on your first treatment. Treat areas mentioned in harborage area using sprays, dust and baits.

Baits - would be the first treatments. This will determine if you only have a few cockroaches that have been brought in or wandered in from the outside.

Sprays and Dust - would only be used if there's an infestation found.

Indoor Treatment

♦ On the inside, apply sprays only in cracks and crevices and as a spot treatment.

♦ Place baits under appliances and according to label directions. Indoor treatments also should consist of all dark, humid or moist areas such as basements, pipes, chases, sewers and around dishwashers.

Outdoor Treatment

♦ Outdoor treatments consist of foundations, crawl spaces, sewers, water meter pits and areas where debris may accumulate.

♦ For Oriental cockroaches, inspect and treat basements, and crawl spaces with extreme dampness. Treat the outside areas around wood, storage, garbage cans and mulch beds. Remove all debris away from the home.

Chemical Recommendations

♦ Max Force - Roach bait and roach gel.
♦ Flushing agent - PT 565 with Pyrethrum.
♦ Liquid Residual - Tempo wettable powder.
♦ Insect Growth Regulator - this will help make them sterile.
♦ Dust - Drione Dust has an affect for up to six to eight months without moisture contact.

Sanitation Program for All Cockroaches

♦ Do not bring paper bags in the home, use plastic.

♦ Keep flour, sugar, corn meal and other grains in plastic containers or other containers with seals.

♦ Keep grease from building up around stoves and on counter tops. This will help eliminate possible feeding areas for other pests -- such as ants.

♦ Wipe counters and appliances down with bleaches or ammonia twice or three times a week.

♦ Mop floors with bleach or ammonia, twice or three times a week.

♦ Wash potatoes and other vegetables before storing or using.

Beetles

Cadelle, Cigarette, Confused Flour and Saw Toothed Grain Beetles

♦   Remove all items from cabinets. Throw away all infested items.

♦   Thoroughly vacuum all dead insects and infested grains.

♦   Clean areas very thoroughly with bleach and scrub. Let it dry. Then, treat crack and crevice areas (i.e., cabinet corners) with a residual insecticide (Dursban). Or, you may use a product recommended for this use. Let it stand for about one to two hours. Then place the food in plastic containers.

♦   Check all food products from store before using.

♦   Don't bring paper bags into the home, use plastic.

Carpet and Furniture Beetles

♦   Find the infestation. Repair or throw away infested material if possible.

♦   Check corners of carpets and furniture for lint piles. Treat infested and other areas with a residual insecticide.

♦   Clean fabric and store them in plastic containers with moth crystals.

♦   Vacuum carpet thoroughly. Throw away vacuum bag after use. Then treat carpet with a residual insecticide. Treat areas around corners in cracks of furniture, and pillows in couches and other chairs.

♦ Treat corners of closets.

♦ Place moth balls or crystals in corners of dresser draws and in corners of closets.

Crickets

Camel Cricket

◆ Treat holes, crack and crevice areas, and the foundation of the structure with a liquid or a dust insecticide.

◆ Treat flower beds, grass and around foundation with Granuals (Dursban or Diazanon Granuals).

◆ Treat basements around doors, windows and baseboards with a liquid insecticide.

Pantry Moths

Almond, Angoumois Grain, and Indian Meal Moths, Meal Worms and Rice Weevils

♦ Remove all items from cabinet. Throw away all infested items.

♦ Thoroughly vacuum all dead insects and infested grain.

♦ Clean areas very thoroughly with bleach and scrub.

♦ Let dry and then treat crack and crevice areas (i.e., cabinet corners) with an residual insecticide (Dursban) or a product recommended for this use.

♦ Check all food products from store before using. Don't bring paper bags into the home, use plastic bags instead.

NOTE: Avoid contaminating food if adult moths are found flying in the home. It is okay to use foggers. Foggers will help, however, they will not get rid of the problem or infestation. You must locate and eliminate infestation and throw away infested products. Use a crack and crevice method along with fogging.

Clothes Moths

Firebrat and Silverfish

♦ Silverfish mostly is found in large numbers, especially in homes that have roofs made of wooden shingles. Treat with an insecticidal dust beneath the shingles. Liquid is okay, however, the liquid does not have a long lasting effect as dust.

♦ Drione Dust, which is an insecticidal dust, has an eliminating affect up to six to eight months depending if there is no moisture contact.

Webbing Clothes Moth

♦ Locate and eliminate moth infestation of food and stored grain products.

♦ Find and destroy infested material.

♦ Infested material can be treated directly with a residual insecticide or by fumigating (i.e., fogging). Plus, good house cleaning, keeping fabrics clean and stored in tight containers with moth balls or moth crystals are also effective methods of treatment.

Fleas

American Dog, Brown Dog and Wood Ticks

Indoor Treatment

♦ Always do a thorough inspection, preferably with a flashlight.

♦ Treat behind baseboards, all cracks and crevice areas, and around door and window trim (frames) with Drione Dust.

♦ Spot treat behind pictures, loose ceiling moldings, drapes, inside furniture and underneath beds and dressers.

♦ Treat corners and all around the carpet in the home or business with Dursban liquid insecticide. For wooden floors, treat cracks with a residual insecticide, using a light fan spray.

♦ Most important, treat all dog and cat bedding and rest areas. After treating pet areas, you can remove or wash in hot water or replace with new bedding. Also, have pets treated simultaneously, when the premises are being treated. Drione Dust is also labeled for the treating of pets. Try not to have pets near pets that may not have been treated to avoid reinfestation.

Outdoor Treatment

♦ Keep the grass and the edges of grass cut to at least four to six inches. This will discourage tick and flea build up. Treat the grass with a liquid residual insecticide (Tempo wettable powder) or Dursban or Diazanon Granules. (Granules work when they come in contact with water -- for example, sprinkling the lawn. The granule releases an insecticide that destroys pests.)

- Granules can be used in flower beds and around bushes. Granules also can be used in flower pots. Make sure pets are inside or away when the lawn is treated. Granules will last about six months.

- Treat dog bedding with a residual insecticide (i.e., Dursban, Dragnet or Safratin). Treat crawl spaces with Drione Dust.

- Always treat 15 feet away from pool areas. In recreational parks, treat along edges of grass, walkways and trails. (Always take special care on family picnics, fishing trips and outings. Always check for ticks and other pests in bags, sleeping gear and yourself.)

- Ticks can be removed by a pair of tweezers. Grab the tick very close to the head if at all possible. This will remove the tick without leaving its head imbedded in the wound. When removed, use any antiseptic on the area.

Fleas

Indoor Treatment

- In bedrooms, remove all items off the floor and in closets.

- If pets have been on the bed or on pillows, remove and wash in hot water after treatment.

- In living and dining rooms with carpet, make sure items are removed from the floor.

- In living rooms, make sure pillows are removed from couches and chairs. Vacuum all carpet before treatments.

- Take all pets to the veterinarian for treatment. Or, treat pets with an insecticidal dust that's labeled for treating pets. Use Drione Dust.

- Treat carpets, chairs, couches with a fan spray pattern with a residual insecticide, that's labeled for treating fleas. Dursban, Dragnet or Safratin is recommended.

- Treat kitchen floors very lightly with a light fan spray pattern. Treat wooden floors also very lightly. Make sure cracks inside wooden floors are treated. Fleas will lay eggs between cracks.

- Treat around and behind baseboards, and pillows on couches and chairs.

- Treat all pet bedding or lay areas moderately heavy. Wash pet bedding in hot water or throw away. Do not soak.

- Flea treatments should be done once or twice indoors per year, depending on the infestation.

- After treating the indoors, leave premises for at least three to four hours. This will allow the chemical time enough to dry. Also, make sure all windows and ventilation systems are off. You want the indoors to be very humid inside. This will help fleas to come out of the larva. Further, this will help to destroy 90% of the adult fleas.

- When returning to premises open windows for good ventilation. Vacuum carpet twice a day for seven days. Throw away vacuum bags after each use. This will keep adult fleas from returning to the carpet.

Outdoor Treatment

- Treat all areas such as cracks in pavements, porches, patios, garages (if pets have spent time in this area), dog houses, kennels, lawns and underneath houses.

- Treat lawns with Dursban or Diazanon Granuals. Granuals will not hurt pets or humans. Granuals release an insecticide when touched by water.

♦ Make sure the grass stays cut low. Fleas will live in shaded areas of the lawn, or underneath trees or bushes and underneath crawl spaces, where dogs and cats go. Fleas will not live in the direct sun light because they will dry out.

♦ Treat underneath crawl spaces with Drione Dust. Drione Dust will also destroy ants, spiders, black widow spiders, crickets and other insects that are labeled pests.

Bed Bugs

♦ Wash bedding in very hot water or throw away. Treat all cracks and crevices in the mattresses. Let mattresses dry for about three hours.

♦ Treat all cracks and other hiding places mentioned above. Use a liquid insecticide for the mattress and also for the cracks in flooring. Use an insecticide dust that is labeled for the use in bed bugs and lice control (preferably, Drione Dust). This process should be done at least one other time, within 15 to 35 days for a follow up treatment. Thoroughness is the key to bedding control.

Lice

♦ For head lice, the hair has to be shaven or cut very, very short. It must also be scrubbed clean with a medication. For the use in treating head lice, depending upon the situation, seek medical attention or talk to a pharmacist.

♦ The same methods should be used for pubic area lice (crabs) and body lice. Body lice can be a very serious problem that needs very special care and attention. This situation is very delicate. In any case, clothes, bedding and other materials would have to be thrown away or burned.

♦ For the treatment of your premises, the following equipment is needed for your protection and others. They are rubber or nylon gloves, hand duster, respirators, spray tanks, goggles, long sleeve shirts and pants. These items should be thrown away after treatment (especially when using chemicals).

Chemical Recommendations

♦ Dursban residual insecticide (spray liquid).
♦ Drione Dust (for crack and crevice use).
♦ You can also use fumigation smokers/foggers by Raid Max.

NOTE: This equipment can be used for every job you do for exterminating pests around your home. The cost of equipment: you can get a metal spray tank at any home improvement store for about $25 to $45 or less. A pair of nylon gloves cost $2, goggles cost about $5 or less, and old long sleeve shirts and pants can be found around the home (wash after use in hot water separately). Hand dusters and respirators, at any pest supply warehouse store, cost about $25 to $40 together.

♦ Treat all crack and crevice areas, behind baseboards and window frames with Drione Dust.

♦ Treat furniture and bed mattresses with a residual spray (treat in the cracks of furniture and mattresses). Also treat corners, seams and around buttons of furniture and mattresses very thoroughly. Let dry before putting on bedding.

♦ Treat wooden floors and carpets with a residual insecticide, using a fan spray pattern. Also, treat corners of carpets. If wooden floors, use a light fan spray pattern. Check the label, some chemicals may stain floors and carpets.

♦ Use an insecticide fogger or fumigating smoker. Leave the room or premises for about three to four hours to give the chemicals time to dry. You can return within two hours to open windows to ventilate. When this is done, clean the mattresses with a mixture of water and bleach and let dry for an hour before putting on the bedding.

Sanitation

♦ Attic Floors - Apply a light coating of Drione Dust on the attic floor.

♦ Basements - Apply Drione Dust in dry areas, such as wall voids (holes in wall) and around plumbing. Use a liquid insecticide, such as Tempo Wettable Powder or Knox-Out (Diazanon) in damp areas. These liquid insecticides suspend themselves in water.

♦ Crawl Spaces - Apply Drione Dust in vent openings. For full coverage use a power duster.

♦ Living Areas - Treat behind baseboards, cracks in walls, behind door and around window frames and cracks around bath tubs and pipes. Also, treat behind and around cabinets, heating units and where pipes leave the premises to the outside. Treat around storage areas in attics, closets, basements and garages. (Baits also can be used to control these pests.)

Don't bring paper bags into the home, use plastic. Don't keep boxes in basements or the attic, use plastic containers.

Rats and Mice

Inspection

- **Droppings** - Fresh droppings of feces are usually moist, soft, shiny and dark. They will become dry and hard in a few days. Old droppings are dull and grayish and will crumble when pressed.

- **Urine** - Dried rodent urine will show a fluorescent color of bluish white to yellowish white. Commercial black lights are often used to detect rodent urine. The use of a black light is not a guarantee that urine is present. You can see other items that will show, when under a black light, such as optical bleaches found in detergent and lubricating oils. For positive identification, use a "brom thymol blue urease test." Place the suspected material on urease-brom thymol blue test paper. Moisten with water and cover with a glass. If a bluish spot appears after three to five minutes, it is rodent urine.

House Mouse, Norway and Roof Rats

Inspection/Sanitation/Exclusion Treatments

Conditions that could offer shelter or food for rodents on the outside and attract rodents to the structure, include the following areas.

- Holes in wall, poorly fitted doorways or other potential rodent entrances.

- Harborage that could be used by rodents, such as crawl spaces, attics or other hiding places.

- Food items that have been eaten or would probably be chosen by rodents.

Good housekeeping and proper storage practices should be used to discourage rodents. These practices will eliminate their access to food and harborage areas.

♦ Keep food from building up on counter tops, appliances, behind dish washers, underneath stoves and behind/under refrigerators.

♦ Keep trash from building up in rooms, basements, garages and around the exterior of the home. Make sure there is no debris, trash or food building up around the exterior of the building. Make sure trash canisters are tightly sealed.

♦ Have all doors, windows and garage doors sealed with strong and thick rubber seals for a tight fit. Remember mice can squeeze through any hole the size of a dime. Rats can squeeze through any hole the size of a quarter. Make sure all door openings are no larger than ¼ of an inch underneath.

♦ Have all pipes around plumbing areas sealed with a foam sealing or metal rings.

♦ Check for holes around the exterior foundation wall. If holes exist, use cement or steel wool.

♦ Make sure lumber is placed on braces, keeping wood six inches off the ground.

♦ Check for moisture around the home and make sure water drains are working properly. Keep the water flowing away from the home or building.

Treatments

Try not to use baits or chemicals as your first and only response. This could be a safety hazard for children and pets.

♦ Use exclusion first -- build them out by rodent proofing.
♦ Use rat and mice traps.

♦   Use baits.

NOTE: Keep the grass cut short. Trim shrubs neatly and keep drainage and gutters clear and free of debris. Rodents are discouraged by surrounding the building foundation with a 18 to 24 inch strip of ⅛ inch pebbled rock in a trench of four inches deep. This is an excellent area for traps and bait stations. Line the trench with roofing or weed screened paper to keep weeds from growing.

Diseases Caused by Rats

♦   *Murine Typhus* - Transmitted by oriental rat flea and prevalent in the south and southeast. A mild *febrile* disease (feverish) that is marked by headaches and rash. It is caused by a *rickettsia*, (rickettsia mooseri) which is a micro-organism.

♦   *Infectious Jaundice or Weill's Disease* - Caused by *spirochete* (bacteria) found in the blood and urine of rats. Contracted by handling or eating things contaminated by rat urine or by swimming in contaminated water.

Mosquitoes

Mosquitoes

♦ Place granules in plants.

♦ Replace screens on windows and doors to ensure a tight fit. Make sure doors have screens. Make sure all windows and screens have 16 to 18 mesh screens. If the window is not in use, seal it off and leave no open cracks.

♦ Make sure there is no standing water in basements and around the home. Make sure all gutters are cleaned at least once a year and all gutter lines (pipes) are working properly and unplugged. Make sure there are no plumbing leaks in and around the home or business.

♦ Make sure trash cans are cleaned at least once a week.

♦ Inspect bird baths and clean at least once a week. Inspect water fountains. Do not treat around fish ponds. Inspect vegetation areas for breeding sites if there is a problem. Use an oil base spray insecticide. Light fan spray over water surface at least twice a month -- once a week is okay -- during the months of April, May, June, July, August and September. Do not treat around fish ponds. Do not treat around water supply for humans and animals, which is used for human consumption.

♦ Use sprays with Pyrethrum in them for the use of spraying vegetation, flowers and lawns. Use foggers for indoor treatments to destroy adult insects.

♦ Use mosquito repellent plants. Place one in the kitchen and in each room. Place one or two on the patio or porch. Also, place one on each corner of the house. Mosquitoes won't come within 100 yards of the plant.

House Fly

♦ First, a good sanitation program is the best and most important method for fly control. Remove all garbage and breeding material for flies. Make sure all trash cans have a tight fitting lid on them. Wash trash cans out at least once a week.

♦ Check screens to make sure there are no holes and they are tightly fitted. Make sure doors have screens on them. Make sure there are no holes around pipes leading into the home.

♦ Make sure dog areas are kept clean and washed down at least once a week. Use a residual spray (Tempo, Knox Out or Diazanon) on the outside of the building. Tempo is best because it will not hurt plants or vegetation.

♦ Inside the home, you can also use Tempo in window seals made of wood, door frames and around baseboards. Knox-Out is good on metal frames. Fogging is good for eliminating 98% of fling insects inside the home.

♦ There are also aerosols and fly time release aerosol mist machines that can be purchased from pest supply warehouses. These machines can be placed in garages, storerooms and in basements. They cannot be used in kitchens, dining rooms or where food is prepared or eaten.

Diseases Caused by Mosquitoes

♦ *Dengue Fever* - a disease of tropical and subtropical regions caused by a virus carried by mosquitoes. The disease comes on suddenly. The patient has a fever, pain in the joints and muscles and a skin rash. Although the original attack usually lasts only two to three days, there is often a relapse on the fourth and fifth days of illness. Treatments consist of bed rest. Aspirin relieves the pain and reduces the fever.

♦ *Filariasis* - caused by parasitic roundworms called *filariae* (singular filaria).

Microscopic young filariae enter the body through the stinger of a mosquito. When they become adults, the filaria blocks the lymph glands, thus preventing fluids from draining from the tissues and causing the limbs to swell. Filariasis is common in the tropics. One of the most dreaded results of the disease is a condition called elephantiasis. Surgery was the only treatment until 1947, when a vaccine was developed to destroy filariae.

♦ *Elephantiasis* - a condition in which parts of the body, usually the legs and external genital organs, become greatly enlarged. The skin of the part involved becomes greatly thickened and lies in folds, so that it resembles the thick hide of an elephant. Treatments consists of pressure bandages, elevation of the affected part, use of drugs, and sometimes surgery. Elephantiasis is most common in the tropics.

♦ *Malaria* - An acute or chronic infectious disease common in tropical and subtropical regions. It is caused by a protozoan, a microscopic animal parasite that attacks the red blood cells. The parasite is transmitted by the female of certain species of "anopheles" mosquitoes. Although rare in the United States, malaria is one of the most prevalent of all diseases. Malaria received its name from the Italian words *"mala aria,"* meaning *"bad air."* Symptoms include aches throughout the body, severe headache and a rapid pulse. After four to six hours or longer, the fever subsides and is followed by sweating and weakness.

♦ *Yellow Fever* - An infectious disease caused by certain kinds of mosquitoes. Yellow fever symptoms usually appear three to six days after the mosquito bite. In the mild, more common form of the disease, the symptoms consist of fever, headache and sometimes jaundice. This is a condition in which the skin turns yellow, caused by excessive bite pigments in the blood. It can also cause a fever with a slow pulse and one suffers severe pains. Yellow fever can also cause internal bleeding. This will cause the patient to vomit dark blood. In fatal cases, death usually occurs from the sixth to the ninth day. Check other sources under yellow jaundice.

Slugs and Snails

Slugs and Snails

♦ Good sanitation is required. This includes cleaning up all dry and damp debris from around the home or building.

♦ Fill in all cracks and holes in bricks and foundation. This will also help with crickets and other pests that hide.

♦ Baits can be purchased which will work, but are very poisonous to pets. A safer and old fashion way would be that you can use beer or garlic. Put beer in a bowl or can. Place it in the ground or on top of the garden or other areas where snails and slugs have been a problem.

Ants

Argentine, Carpenter Bee, Harvester, Honey, Little Black, Pharoah, Southern Fire and Thief Ants

Indoor Treatment

♦ First do a thorough inspection using a flashlight with good batteries. If the nest is in walls, hollow doors, in floors and pipes, apply a residual insecticide dust (Drione Dust).

♦ It is okay to use liquid, however, a liquid only has a surface contact residual. It is not able to get into all areas that a dust can. Drione Dust has Pyrethrum -- an insecticide with a fogging agent in it. The dust will not only destroy insects and pests where they live, but will penetrate other areas. (When using a residual liquid insecticide/pesticide, use it around baseboards, corners, doors and window seals. Liquids can only be used for crack and crevice and spot treatment unless otherwise stated on the label.)

♦ Baits - could be used first as a monitoring treatment to eliminate ants. Keep baits out of the reach of children and pets. Do not contaminate food surfaces. If ants are in heat ducts or vents use baits.

Outdoor Treatment

♦ Same as indoors - do a thorough inspection as well as locate the nest. Apply liquid insecticide in and around the mound. Granules also can be sprinkled around flower beds on grass and around homes. When granules are wet, they release the insecticide chemical.

♦ A thorough inspection must be done, looking both outside and inside. Look from the top of the roof or building to the basement of a home or crawl space to foundations of buildings. Follow trails, looking in both directions, where ants are going and where ants are coming from. When ants are moving in a stream, this could indicate that the colony is moving, not looking for food. Individual ants are hunting for food or water.

Carpenter Ant

♦ First locate the nest and look for a trail of ants. Look for sawdust-like material. Look for flying ant wings in window seals. The entire nest must be located and destroyed.

♦ Moisture conditions should be eliminated.

♦ Remove and destroy all decayed wood. Prune (cut) back all tree limbs and bushes away from home. Keep wood piles away from home and at least six inches off the ground.

♦ Use Drione Dust. Apply it to the inside of the nest. The Dust will travel throughout the nest and galleries and destroy the ants.

♦ Treat attics and crawl spaces with a light application.

♦ For outside perimeter treatment use a wettable powder such as Tempo. Tempo will not hurt plants, bushes and vegetables.

Carpenter Bee

♦ The first and foremost step is to do a thorough inspection.

♦ Check out the situation first and wear the proper safety equipment. This includes goggles and a cap, long sleeved shirt, long pants, rubber gloves, hand duster and a respirator if the label recommends.

◆ Dust each hole with a residual insecticide called Drione Dust.

◆ Dust the entrance of the nests by blowing the Dust down into the gallery or holes.

◆ Liquid insecticides are also effective.

◆ Seal off all openings after treating. Plug holes with putty or other plug like materials.

Pharoah Ants

◆ The most important factor is maintaining a good sanitation program. Sanitation is most important in any pest or non pest situation. Treatment for Pharoah ants are very difficult and a very hard and prolonged process.

◆ Pre-bait by using a mixture of water and corn syrup. Place it in a small container. This process should be done about two weeks. Plenty of time will then be allowed for where feeding areas are located and increasing. You can also use pieces of liver for feeding and tracking.

◆ Place baits for Pharoah ants, preferably Max-Force Pharoah ant baits, where ant trails have been found. Pharoah ant baits are the best way to get rid of Pharoah ants.

◆ Spot treatments can be applied at entry points with a wettable powder. Check labels for the use on Pharoah ants.

Bees

Bumble, Honey, Queen and Drone, and Worker Bees, Paper Nest, Solitary Wasps, Yellow Jackets and Hornets

♦ Drione Dust is the best and most effective for this type of job. Because of the danger of their sting, the best time to treat or remove a nest is at night -- when all the wasps are in the nest and it is cool or in the early morning before daylight. Do not ever treat when the sun is shining because bees will sting you.

♦ The first and foremost step is to do a thorough inspection. Check out the situation first and wear proper safety equipment -- goggles, a cap, long sleeved shirt, rubber gloves, hand duster and long pants. Do not treat if you are allergic to stings. Have someone else treat who is not allergic.

♦ Use a Pyrethrum aerosol to stun the wasps and then apply a heavy coating of dust or spray. When applying these chemicals, force them into the entrance of the hole in the nest.

♦ Once insects are destroyed, knock the nest down. Destroy it by crushing or burning it. Do not burn the nest if it is in a tree or bush. It will damage the plant.

NOTE: When stung by a bee or wasp try to remain calm. Keep yourself from getting nervous or going into a panic situation. You can use the old fashion method of taking tobacco from a cigarette or cigar and then applying moisture (water or saliva) to the tobacco. Place the mixture on the wound. This will bring out the stinger and help ease the pain. Do not try this if you are allergic to bee and wasp stings. Seek medical attention as quickly as possible.

Termites

Damp-Wood, Dry-Wood and Subterranean Termites

There are two ways you can treat for termites. They are as follows.

♦ Learn about the termite - how they live, what they look like (description), food habits, work habits, where they live, the description of the damage they cause, how to treat (exterminate) them, what chemicals to use and where to buy them.

♦ Have a licensed exterminating company to do it for you. Be sure to verify and to see if the job is done right.

Recommendation from the Department of Agriculture

This should be used by builders/contractors and people buying a home or having a home built. The consumer should use stones, bricks or concrete for the foundations of the home. Bridges, homes, trestles, and support posts of buildings should also be made out of stones, bricks or concrete. If timber is to be used, make sure it has been treated with Creosote. Creosote is an oily liquid distilled from wood tar or coal tar. It is used as an antiseptic and a wood preservative. Or, you can use pressure treated wood. Most termites cannot live without water. Cutting off their water supply or stopping moisture and any dampness from water supply will usually kill termites.

Spiders

Banded Garden, Black Widow, Brown Recluse, Grass, House, Jumping, Orange Garden, Tarantula, Trap-Door Tarantula and Wolf Spiders

Spider treatments can be done at anytime. The preferred time is during the day. Daylight will help you spot spiders more easily. Otherwise, some spiders stay hidden until disturbed.

Indoor Treatment

♦ The first and foremost step is to do a thorough inspection.

♦ Use a crack and crevice treatment with a liquid or dust insecticide (preferably Drione Dust). Dust should be used only indoors for crack and crevice treatment for spiders. You can use a liquid insecticide for spot treatments and around baseboards.

♦ Vacuum webs and eggs from corners and other places. After a thorough vacuuming, throw away the vacuum bag.

♦ Treat behind and underneath furniture, appliances (i.e., refrigerators and stoves), cracks around windows, inside window seals, corners of closets, kitchens, bathroom sinks, cabinets and curtains.

Outdoor Treatment

♦ Check sanitation to make sure there is no trash, boxes, wood or other debris around the home. If there is, clean and remove debris. This will also help eliminate hiding places for spiders and other pests.

♦ Check under shelves, behind window shutters and corner of doors and windows to see if there are spiders or spider webs. Vacuum or knock down with a duster. Then, treat the area with an liquid insecticide (labeled for the use on spiders).

♦ If the home has a crawl space, dust with Drione Dust. Use a hand or blow duster.

♦ Check corners of roofs, patios, sheds and storage spaces. Treat appropriately.

NOTE: Never treat any area without wearing the proper safety equipment. Remember to always read the label.

Common Household Insecticides and Their Uses

Drione Dust is an insecticide that is a *desiccant dust*, known as a drying agent. It provides "quick control" and kills pests up to six months when left undisturbed. That is, when it is left dry and with no moisture contact.

This product will kill mostly all pests. It can be placed in crack and crevice areas such as wall voids and in walls where kitchen and bathroom plumbing are found. All these areas are where cockroaches and ants travel. Another good place to use this insecticide is inside *bath traps*. A bath trap is a panel located in closets or rooms where the back of the shower and tub plumbing can be reached.

Drione Dust releases a fogging agent within the dust called *Pyrethrum*. Pyrethrum is an insecticide that fogs the air for three to four hours. The consumer would call it a bomb. Pyrethrum or fogging has no residual effects and no chemicals to contact surfaces.

The bombs the consumer buys from the store allow cockroaches and other pests to hide until bombs disappear -- they only stay in the air. These products are called *flushing agents*. This allows German cockroaches to use their antennas to detect insecticides in the air. Therefore, this will not stop cockroaches and other pests from having full contact with the insecticide. However, Drione Dust flushes and kills simultaneously.

Buying these kinds of flushing agents (bombs) is not worthwhile. They only defeat the purpose, unless the inside walls have been treated first. Drione Dust is one of the most effective tools in getting rid of problem pests, described in the following list.

- Ants
- Bedbugs
- Bees
- Box Elder Bugs
- Cadelles
- Cheese Mites
- Cigarette Beetles
- Cockroaches
- Confused Flour Beetles
- Crickets
- Drywood Termites
- Firebrats
- Fleas
- Lice
- Rice Weevils
- Saw-toothed Grain Weevils
- Silverfish
- Spiders
- Ticks
- Wasps
- Yellow Mealworms

These treatments must be done in a "crack and crevice" manner. This will aide in controling these pests.

Cautionary Procedures to Use Before Signing Pest Control Company Contracts

Before signing an agreement with the pest control company ask to see a license. Ask the technician if he or she is licensed and has a license. If the technician has a license they will probably know what you are talking about. If they do not, then there is a good chance that the technician has not completed his or her training. Ask the pest control company for someone who is licensed.

Before hiring a pest control company, try to remedy the problem yourself. For example, determine what type of pests you have. Ask yourself, "have we seen these many ants before?" First, purchase a can of bug spray before calling the pest control company, because no one is more familiar with this than you are. The pest control company is going to rely upon the consumer to give it this type of information. Afterwards, they will only inspect and inform the consumer of their inspection results. The consumer is more knowledgeable about the existence of these pests. Thus, the pest control company and its inspectors are working for the consumer.

If any pest company comes out to your home or business and tries to force you to sign a contract, ask them

to leave. A consumer does not need this type of hassle. It
is unnecessary.

If a pest control contract already exists, there are
certain things you need to know. For example, if your
pest control company is late or does not do their job
correctly, they have violated the contract. If they fail to
meet their obligation as agreed upon, the consumer has the
right to end the contract. Pest control companies do not
want the consumer to know this because they will lose
money if you cancel your contract. That is why they
hassle you. Further, they want the consumer to think that
the contract cannot be canceled. Pest control contracts,
like any others, are enforceable. If it is violated, the
consumer has the right to cancel the contract.

Example 1 - Mrs. McNaps calls a pest control
company for ants. She agrees to sign
a one year contract. The company
says: they will be on time, do the job
right the first time or do it over, and
provide maintenance services according to Mrs. McNaps' schedule. If
any of these terms are neglected,
Mrs. McNaps is not under any obligation to stay in the contract. Therefore, the contract becomes *VOID*.

Example 2 - Let's say there are ants in your
kitchen. You have tried to get rid of
them and failed. The pest control
company is called, who sends out an
inspector. The inspector agrees there
are ants and several treatments are
needed to get rid of the problem.

Because the problem is annoying, the first inclination is to trust the inspector. So, a one-year contract is signed. Also, the consumer pays the initial (first time) service cost of $96.00-$120.00 dollars. What is agreed to, is a monthly obligation of $25.00-$60.00 to get rid of the ants. Question for the consumer, does it take the pest control company a year to get rid of ants?

After the pest control company comes out to perform its services, check to see whether the pest control company has done its job properly. The consumer should check to see that there are no dead bugs and smell of pesticides. Then, call the company to verify that someone came out to your home.

After the contract ends after a year, are you, the consumer, educated on how to keep these pests from coming back? Is there any educational program offered to you? Has anyone offered to educate the consumer on how and where to buy or use insecticide products?

Again, this is not to imply that all pest control companies are like this. There are some companies that really care about the needs of the people. This book is not an effort to downgrade or slander pest control companies. It is written to help the consumer find a very honest pest control company. One, which will care enough to educate the consumer about their pest control needs.

NEVER SIGN A ONE YEAR CONTRACT WITHOUT CALLING AN EDUCATIONAL CENTER OR THE DEPARTMENT OF AGRICULTURE. ASK QUESTIONS AND GET ANSWERS, FIRST, ABOUT ANY PEST PROBLEMS.

Residential Pest Treatment Agreement
One or More Treatments

COMPANY

_____

_____

_____

PREMISES TO BE TREATED

Address_____

City_____

State_____Zip_____

Phone (Home)_____

Phone (Work)_____

Inspection Date_____ Date of Treatment_____ Follow-up Date_____

Follow-up Treatment (No Charge)

1 Follow-up Treatment__ 2 Follow-up Treatments___ 3 Follow-up Treatments__

Guarantee: None_____ 30 Days_____ 60 Days_____ 90 Days_____

Treatments: One Time__ Weekly___ Bi-Weekly___ Monthly___ Bi-Monthly__ Quarterly___

Special

Children   Yes____ No____   Pets   Yes____ No____   Kind of Pet(s) _____

New Born   Yes____ No____   Anyone Pregnant   Yes____ No____

Senior Citizen   Yes____ No____   Anyone Allergic   Yes____ No____

Any Oxygen   Yes____ No____   Anyone Sick   Yes____ No____

Any Special Areas Need Not Be Treated (At Customer Request)   Yes____ No____

Where:_____

Pest(s)

Cockroaches   German____   American____   Oriental____   Brown Banded____

Ants   Carpenter____ Little black ant____ Thief Ant____ Fire Ant____ Pharaoh_____

  Other_____

Spiders   Black Widow_____ Brown Recluse_____ Wolf_____ House_____

  Other_____

Rats   Norway_____ Roof_____ Mice_____ Crickets_____ Centipedes_____

  Millipedes_____ SilverFish_____ FireBrat_____ Earwigs_____ Fleas_____

  Bees_____ Wasps_____ (We will treat and destroy nest only)

NOTE: Outside flea treatments (will extend 10 feet from premises) (Lawn care flea treatments are an additional charge).

Residential Pest Treatment Agreement
One or More Treatments (Continued)

Paid by: (Home Owner)____ (Renter)_____ (Other)_____

Treatment Cost...................... $_____   Cash_____

Senior Citizens Discount 15%..._____   Money Oder_____

Referral Discount 10%.............._____   Credit Card #_____

Tax...____   Exp. Date._____Driver Lic. #_____

Total...____

_____   _____

Company Rep.   *Date*   *Customer Acceptance*   *Date*

Sample

Commercial One Time or More
Pest Treatment Agreement

COMPANY

PREMISES TO BE TREATED
Address_____
City_____
State_____Zip_____
Phone_____
Contact Person _____

3 Months or More Treatments

Inspection Date_____ Date of Treatment_____ Night_____ Date_____

Regular Services: Night_____ Date_____ Weekday_____ Saturdays_____
                Day of Week_____ Time_____

Extra Services: 2 Per Month

Treatments: Weekly_____ Bi-Weekly_____ Monthly_____ Bi-Monthly_____
               Quarterly: Every 3 Months_____ Every 4 Months_____

One Time Treatments

Inspection Date_____   Date of Treatment_____
Follow up Date_____ (No Charge) None____ 1_____ 2_____ 3_____
Guarantee: None_____ 30 Days_____ 60 Days_____ 90 Days_____

Special

Is there any one pregnant? Yes___No____   Are there any sick? Yes___No____
Are there any persons allergic to pesticides? Yes___No____

Commercial One Time or More
Pest Treatment Agreement (Continued)

Pest(s):	Cockroaches	Centipedes	Millipedes	Earwigs	Crickets
	Rats	Mice	Spiders	Ants	Fleas
	Bees	Wasps	Special_____		

Initial Treatment Cost $_____   Per Treatment Cost $_____

_____   _____
Company Rep.   Date   Customer Acceptance   Date

3 Days Cancellation

Customer has thee days to cancel this contract after signing, before the service/treatment starts. After three days, the customer agrees to accept the contract.

_____   _____
Company Rep.   Date   Customer Acceptance   Date

30 Days Cancellation

Customer may cancel the contract, by giving a 30 days' advance notice before the next month's service. After the 30 days' notice, the customer will no longer be obligated to keep the contract.

Customer must notify company in writting, by sending letter by mail, having the postmark (reflecting the 30 days') or by hand delivery to office or by technician.

Appendix

Appendix A - Retail Outlets

Special Warehouses

For Mosquito Repellent Plants: Michigan Bulb Company
1950 Waldorf, NW
Grand Rapids, MI 49550
Phone: (616) 771-9500

For Beneficial Insects: Gurney's Seed and Nursery Company
110 Capital Street
Yankton, South Dakota 57079
Phone: (605)665-1671

For Pesticides/Insecticides: York Distributors - 303
5185 Raynor Avenue
Linthicum Heights, MD 21090
Phone: (410) 636-2400 or (800) 235-6138

Appendix B - Agency and Internet Resource Listings

Information Sources on Agencies Supporting Pesticides and Pest Control

♦ *Bio-Integral Resource Center* (BIRC) has information on least toxic methods for pest management. You can contact them by writing to P.O. Box 7414, Berkeley, California 84707.

♦ *California Department of Pesticide Regulation's Environmental Monitoring and Pest Management Branch* has published a booklet titled, *Suppliers of Beneficial Organisms in North America,* on biological control organisms. Copies of this booklet are available by writing the Department at 1020 N Street, Room 161, Sacramento, California 95814-5624. The telephone number is 916-324-4100.

♦ *County Cooperative Extensive Offices* are affiliated with the Land Grant university in each state. It is usually listed in the telephone directory under county or state government. These offices often have a range of resources on lawn care and landscape maintenance, including plant selection, pest control, and soil testing.

♦ The state pesticide regulatory agencies are responsible for enforcement of pesticide laws and certification of pesticide applicators, according to the cooperative agreements. These agencies seek to provide programs to protect agricultural workers, prevent pollution of ground water, conserve threatened and endangered species, promote integrated pest management, and reduce the risks of pesticide use. Complaints from citizens that involve the application of pesticides are referred to the appropriate state agencies. They also provide answers to questions and information about pesticide laws and regulations, pesticide safety, and the proper use of pesticides. Pesticide product registration kits may be obtained by calling *EPA's Registration Division* at (703) 308-8341.

♦ *National Antimicrobial Information Network* (NAIN) provides objective information to callers regarding antimicrobial products. Inquiry calls are welcome and the service is open for everyone's use. NAIN receives reports of antimicrobial product problems, ineffectiveness and incidents. This information is reported to the EPA, and is used by the Agency in formulating regulatory policies. NAIN can:

♦ help callers with interpretation of product labels and permitted uses.

♦ refer requests that are outside the expertise or authority of NAIN to more appropriate sources.

♦ refer callers for injury to humans or animals and laboratory analyses.

♦ supply general information on regulation of antimicrobials in the United States and lists of products registered with the EPA.

♦ provide safety information, health and environmental effects.

♦ take reports of lack of product efficacy and forward this information to the EPA.

NAIN is a service that provides objective, science-based information about a wide variety of antimicrobial-related subjects, including:

♦ antimicrobial products
 ♦ sanitizers
 ♦ disinfectants
 ♦ sterilants
 ♦ toxicology
♦ environmental chemistry.

NAIN is a toll-free telephone service operated in conjunction with the National Pesticide Telecommunications Network. This service provides antimicrobial information to any caller in the United States, Puerto Rico, or the Virgin Islands. NAIN's operating hours are 7:30 a.m. to 4:30 p.m., Pacific time, Monday through Friday, excluding holidays. The telephone number is 1-800-447-6349. The fax number is 1-541-737-0761.

♦ *National Center for Environmental Publications and Information* provides a listing of hundreds of pesticide publications, including science chapters and fact sheets and other information media. To order write: P.O. Box 42419, Cincinnati, OH 45242-2419. Telephone: 1-800-490-9198 or 513-891-6561; Fax 513-891-6685.

♦ *The National Pest Control Association* has various publications on pest control and the pest control industry. They publish Pest Fact Sheets that are available to the public. Write to: 8100 Oak Street, Dunn Loring, Virginia 22027. Telephone: (703) 573-8830. Fax: (703) 573-4116.

♦ *The National Pesticide Telecommunications Network* (NPTN) - 1-800-858-7378 (general public), 6:30 a.m.-4:30 p.m. Pacific Time (9:30 a.m.-7:30 p.m., Eastern time) Monday-Friday. NPTN is a national pesticide information service funded through the EPA Office of Pesticide Programs. NPTN information specialists provide information on pesticide products, safety, health effects, environmental effects, clean-up procedures and disposal. They also provide referrals for human or animal poisonings, laboratory analysis and pesticide incidences. NPTN is located at Oregon State University in Corvallis, Oregon.

NPTN is a toll-free service that provides pesticide information to any caller in the United States, Puerto Rico, or the Virgin Islands. It is a service that is staffed by highly qualified and trained pesticide specialists who have the toxicology and environmental chemistry education and training needed to provide knowledgeable answers to pesticide questions. A source of factual chemical, health, and environmental information, NPTN provides information on about the more than 600 pesticide active ingredients incorporated into over 50,000 different products registered for use in the United States since 1947.

NPTN can help callers interpret and understand toxicology and environmental chemistry information about pesticides. NPTN can access over 300 pesticide resources, pesticide label information and supply general information on regulation pesticides in the United States. Callers to NPTN are directed to the appropriate persons for pesticide incident investigation and emergency human and animal treatment, along with others.

♦ *PCT Online* - the pest control industry's most innovative interactive website. Features on this website includes monthly features of PCT and service technician magazines online, late-breaking news, on-line subscription service, links to other industry websites, on-line order forms for PCT's complete technical library, PCT dialogue business and technology conference updates, classified advertising, PCT commercial product guides on-line, cutting edge market research and exclusive service technician training supplements. Website address is: www.pctonline.com.

♦ *Pesticide Information Network* (PIN) is an interactive database system containing current and historic pesticide information. It is free and operational 24 hours a day, seven days a week. It can be reached via modem and communications software at 703-305-5919.

♦ *State agriculture agencies and/or environmental agencies* may publish information on pests and pest management strategies. The state pesticide regulatory agency can provide information on pesticide regulations, and may also have information on companies with a history of complaints or violations. NPTN can identify the agency responsible for pesticide regulation in each state.

Environmental Protection Agency Addresses

Headquarters

U.S. Environmental Protection Agency
Office of Pesticide Programs (7506C)
401 M Street, S.W.
Washington, D.C. 20460
Telephone: (703) 305-5017
Fax: (703) 305-5558

EPA Regional Offices

U.S. Environmental Protection Agency,
  Region 1
Air, Pesticides and Toxic Management
  Division
State Assistance Office (ASO)
1 Congress Street
Boston, MA 02203
Telephone: (617) 565-3932
Fax: (617) 565-4939

U.S. Environmental Protection Agency,
  Region 2
Building 10 (MS-105)
Pesticides and Toxics Branch
2890 Woodbridge Avenue
Edison, NJ 08837-3679
Telephone: (908) 321-6765
Fax: (908) 321-6788

U.S. Environmental Protection Agency,
  Region 3
Toxics and Pesticides Branch (3AT-30)
841 Chestnut Building
Philadelphia, PA 19107
Telephone: (215) 597-8598
Fax: (215) 597-3156

U.S. Environmental Protection Agency,
  Region 4
Pesticides and Toxics Branch (4-APT-MD)
345 Courtland Street, N.E.
Atlanta, GA 30365
Telephone: (404) 347-5201
Fax: (404) 347-5056

U.S. Environmental Protection Agency,
  Region 5
Pesticides and Toxics Branch (SP-14J)
77 West Jackson Boulevard
Chicago, IL 60604
Telephone: (312) 886-6006
Fax: (312) 353-4342

U.S. Environmental Protection Agency,
  Region 6
Pesticides and Toxics Branch (6PD-P)
1445 Ross Avenue
Dallas, TX 75202-2733
Telephone: (214) 665-7240
Fax: (214) 665-7263

U.S. Environmental Protection Agency,
  Region 7
Water, Wetlands and Pesticides Division
726 Minnesota Avenue
Kansas City, KS 66101
Telephone: (913) 551-7030
Fax: (913) 551-7065

U.S. Environmental Protection Agency,
  Region 8
Air, Radiation and Toxics Division (8ART)
One Denver Place, Suite 500
999 18th Street
Denver, CO 80202-2405
Telephone: (303) 293-1730
Fax: (303) 293-1229

U.S. Environmental Protection Agency,
  Region 9
Pesticides and Toxics Branch (A-4)
75 Hawthorne Street
San Francisco, CA 94105
Telephone: (415) 744-1090
Fax: (415) 744-1073

U.S. Environmental Protection Agency,
  Region 10
Pesticides and Toxics Branch (AT-083)
1200 Sixth Avenue
Seattle, WA 98101
Telephone: (206) 553-1091
Fax: (206) 553-8338

Addresses for State Pesticide Agencies

Alabama
Director
Division of Pesticide Management
Department of Agriculture and Industries
P.O. Box 3336
Montgomery, AL 36109-0336
(334) 242-2656

Alaska
Director
Division of Environmental Health
Department of Environmental Conservation
410 Willoughby Avenue, Room 107
Juneau, AK 99801-1795
(907) 465-5280

America Samoa EPA
Office of the Governor
American Samoa Government
P.O. Box 2609
Pago Pago, American Samoa 97699
(684) 633-2304

Arizona
Director, Environmental Services Division
Arizona Department of Agriculture
1688 West Adams
Phoenix, AZ 85007
(602) 542-3578

Arkansas
Director, Division of Pesticides
Arkansas State Plant Board
#1 Natural Resources Drive
Little, AR 72205
(501) 225-1598

California
Director
California Department of Pesticide Regulation
1020 N Street, Room 100
Sacramento, CA 95814-5624
(916) 442-4300

Colorado
Director, Division of Plant Industry
Colorado Department of Agriculture
700 Kipling Street, Suite 4000
Lakewood, CO 80215-5894
(303) 239-4140

Connecticut
Director, Pesticide Management Division
Department of Environmental Protection
79 Elm Street
Hartford, CT 06106
(203) 424-3369

Commonwealth-Northern Mariana Islands
Department of Public Works
Division of Environmental Quality
Commonwealth-Northern Mariana Islands
P.O. Box 1304
Saipan, Mariana Islands 96950
(670) 234-6984

Delaware
Deputy Secretary
Delaware Department of Agriculture
Division of Consumer Protection
2320 South DuPont Highway
Dover, DE 19901
(302) 739-4811

District of Columbia
Program Manager
Pesticide Hazardous Waste and Underground
  Storage Tank Division
Environmental Regulation Administration
Department of Consumer/Regulatory Affairs
2100 Martin Luther King Ave. SE, Rm 203
Washington, D.C. 20020
(202) 645-6080

Florida
Director, Division of Agricultural
  Environmental Services
Department of Agriculture
3125 Conner Boulevard
Tallahassee, FL 32399-1650
(904) 488-3731

Georgia
Assistant Commissioner
Plant Industry Division
Georgia Department of Agriculture
19 Martin Luther King Drive, S.W.
Atlanta, GA 30334
(404) 656-4958

Guam
Pesticide Program Director
Guam Environmental Protection Agency
P.O. Box 22439-GMF
Barrigada, GU 96921
(671) 472-8863

Hawaii
Administrator
Pesticide Programs
Hawaii Department of Agriculture
P.O. Box 22159
Honolulu, HI 96823-2159
(808) 973-9401

Idaho
Administrator
Division of Agricultural Technology
Idaho Department of Agriculture
P.O. Box 790
Boise, ID 83701-0790
(208) 334-3550

Illinois
Chief
Bureau of Environmental Programs
Illinois Department of Agriculture
P.O. Box 19281
Springfield, IL 62794-9281
(217) 785-2427

Indiana
Pesticide Administrator
Office of the Indiana State Chemist
1154 Biochemistry Building
Purdue University
West Lafayette, IN 47907-1154
(317) 494-1585

Iowa
Chief
Pesticide Bureau
Iowa Department of Agriculture
Henry A. Wallace Building
East 9th Street and Grand Avenue
Des Moines, IA 50319
(515) 281-8591

Kansas
Director
Plant Health Division
Kansas Department of Agriculture
109 S.W. 9th Street
Topeka, KS 66612-1281
(913) 296-2263

Kentucky
Director, Division of Pesticides
Kentucky Department of Agriculture
100 Fair Oaks Lane
Frankfort, KY 40601
(502) 564-7274

Louisiana
Director
Pesticide and Environmental Programs
Department of Agriculture and Forestry
P.O. Box 3596
Baton Rouge, LA 70821-3596
(504) 925-3763

Maine
Director
Board of Pesticide Control
Maine Department of Agriculture
State House Station #28
Augusta, ME 04333
(307) 287-2731

Maryland
Chief
Pesticide Regulation Section
Office of Plant Industries and Pest
  Management
Maryland State Department of Agriculture
50 Harry S. Truman Parkway
Annapolis, MD 21401-7080
(410) 841-5710

Massachusetts
Chief
Pesticides Bureau
Massachusetts Department of Food and
Agriculture
100 Cambridge Street, 21st Floor
Boston, MA 02202
(617) 727-3000

Michigan
Director
Pesticide and Plant Management Division
Michigan Department of Agriculture
P.O. Box 30017
Lansing, MI 48909
(517) 373-1087

Minnesota
Director
Division of Agronomy Services
Minnesota Department of Agriculture
90 West Plato Boulevard
St. Paul, MN 55107
(612) 296-5639

Mississippi
Director
Bureau of Plant Industry
Mississippi Department of Agriculture
  and Commerce
P.O. Box 5207
Mississippi State, MS 39762
(601) 325-3390

Missouri
Director
Bureau of Pesticide Control
Missouri Department of Agriculture
P.O. Box 630
Jefferson City, MO 65102
(314) 751-2462

Montana
Administrator
Agricultural Sciences Division
Montana Department of Agriculture
P.O. Box 200201
Helena, MT 59620-0201
(406) 444-2944

Nebraska
Director, Bureau of Plant Industry
Nebraska Department of Agriculture
301 Centennial Mall
P.O. Box 94756
Lincoln, NE 68509
(402) 471-2394

Nevada
Director, Bureau of Plant Industry
Nevada Division of Agriculture
350 Capitol Hill Avenue
Reno, NV 89520
(702) 688-1180

New Mexico
Chief, Bureau of Pesticide Management
Division of Agricultural and Environmental
Services
New Mexico State Department of Agriculture
P.O. Box 3005, Department 3AQ
New Mexico State University
Las Cruces, NM 88003-0005
(505) 646-2133

New Hampshire
Director
Division of Pesticide Control
New Hampshire Department of Agriculture,
  Markets and Food
P.O. Box 2042
Concord, NH 03302-2042
(603) 271-3550

New Jersey
Assistant Director
Pesticide Control Program
Department of Environmental Protection
CN 411
Trenton, NJ 08625-0411
(609) 530-4011

New York
Chief
Bureau of Pesticides and Radiation
Division of Solid and Hazardous
  Materials Regulation
Department of Environmental Conservation
50 Wolf Road
Albany, NY 12233-7254
(518) 457-7482

North Carolina
Assistant Pesticide Administrator
Food and Drug Protection Division
North Carolina Department of Agriculture
P.O. Box 27647
Raleigh, NC 27611-0647
(919) 733-3556

North Dakota
Director, Pesticide Division
North Dakota Department of Agriculture
State Capitol, 600 East Boulevard, 6th Floor
Bismarck, ND 58505-0020
(701) 328-4756

Ohio
Specialist in Charge of Pesticide Regulation
Division of Plant Industry
Ohio Department of Agriculture
8995 East Main Street
Reynoldsburg, OH 43068-3399
(614) 728-6987

Oklahoma
Director
Department of Environmental Quality
Plant Industry and Consumer Services
Oklahoma Department of Agriculture
2800 North Lincoln Boulevard
Oklahoma City, OK 73105-4298
(405) 271-1400

Oregon
Administrator, Plant Division
Oregon Department of Agriculture
635 Capitol Street, N.E.
Salem, OR 97310-0110
(503) 986-4635

Pennsylvania
Chief
Agronomic Services Division
Bureau of Plant Industry
Pennsylvania Department of Agriculture
2301 North Cameron Street
Harrisburg, PA 17110-9408
(717) 787-4843

Puerto Rico
Director
Analysis and Registration of Agricultural
  Materials
Puerto Rico Department of Agriculture
Agrological Laboratory
P.O. Box 10163
Santurce, PR 00908
(809) 796-1735

Rhode Island
Chief
Division of Agriculture
Rhode Island Department of Environmental
  Management
22 Hayes Street
Providence, RI 02908
(401) 277-2782

South Carolina
Department Head
Department of Pesticide Regulation
257 Poole Agriculture Center
Clemson University
Clemson, SC 29634-0394

South Dakota
Administrator
Office of Agronomy Services
Agricultural Services
South Dakota Department of Agriculture
523 E. Capitol, Foss Building
Pierre, SD 57501-3182
(605) 773-4432

Tennessee
Director
Plant Industries Division
Tennessee Department of Agriculture
P.O. Box 40627
Nashville, TN 37204
(615) 360-0130

Texas
Assistant Commissioner for Pesticides
Texas Department of Agriculture
P.O. Box 12847
Austin, TX 78711
(512) 463-7624

Utah
Director
Division of Plant Industry
Utah Department of Agriculture
Box 146500
Salt Lake City, UT 84114-6500
(801) 538-7180

Vermont
Director
Plant Industry, Laboratory and Standards
Division
Vermont Department of Agriculture
116 State Street
Montpelier, VT 05602
(802) 828-2431

Virgin Islands
Pesticide Program Director
8000 Nisky Center, Suite 231
Estate Nisky, Charlotte Amalie
St. Thomas, U.S. VI 00802
(809) 774-3320, ext. 135

Virginia
Program Manager
Office of Pesticide Services
Virginia Department of Agriculture and
Consumer Service
P.O. Box 1163
Richmond, VA 23209
(804) 371-6558

Washington
Assistant Director
Pesticide Management Division
Washington State Department of Agriculture
P.O. Box 42560
Olympia, WA 98504-2560
(360) 902-2010

West Virginia
Director
Pesticide Division
West Virginia Department of Agriculture
1900 Kanawha Boulevard, East
Charleston, WV 25305-0190
(304) 558-2209

Wisconsin
Administrator
Agricultural Resources Management Division
Wisconsin Department of Agriculture
Trade and Consumer Protection
2811 Agriculture Drive
Madison, WI 53704
(608) 224-4546

Wyoming
Director
Technical Services
Wyoming Department of Agriculture
2219 Carey Avenue
Cheyenne, WY 82002-0100
(307) 777-6590

Office of Pesticide Programs, Environmental Protection Agency

Communications Branch (7506C)
Office of Pesticide Programs
U.S. Environmental Protection Agency
401 M Street, S.W.
Washington, D.C. 20460
(703) 305-5017

Biological and Economic Analysis Division - Assessment of pesticide use and benefits; operation of analytical chemistry and antimicrobial testing laboratories. *Allen L. Jennings, Director*

Biopesticides and Pollution Prevention Division - Risk/benefit assessment and risk management functions, biochemical pesticides and plant pesticides. *Janet L. Anderson, Acting Director*

Field and External Affairs Division - Program policies and regulations, legislation and Congressional interaction, general food safety issues, ground water activities, publications and communications, endangered species and worker protection, certification and training. *Anne E. Lindsay, Director*

Information Resources and Services Division - Information support, records, computer support, FIFRA Section 6(a)(2) issues, pesticide incident monitoring, and the National Pesticides Telecommunications Network. *Linda Travers, Director*

Registration Division - Product registrations, amendments, reregistrations, tolerances, experimental use permits, and emergency exemptions for all pesticides not assigned to the Biopesticides and Pollution Prevention Division. *Stephen L. Johnson, Director*

Internet Resources

Agency	World Wide Web Site	Source of Information
Environmental Protection Agency	www.epa.gov	Information on this entire governmental agency
Office of Pesticide Programs (OPP), Division of the Environmental Protection Agency	www.epa.gov/pesticides	All Federal Register publications and press announcements, reregistration eligibility decisions (REDs), information on Food Quality Protection Act (FQPA) implementation efforts and fact sheets and publications of general interest, such as Citizens to Pest Control and Pesticide Safety, and the Catalog of OPP publications.
Oklahoma Department of Agriculture	www.oklaosf.state.ok.us/~okag/aghome.html	Pesticide enforcement under the Federal Insecticide Fungicide Rodenticide Act (FIFRA)
Texas Department of Agriculture	www.agr.state.tx.us/pesticide/index.htm	Pesticide enforcement under the Federal Insecticide Fungicide Rodenticide Act (FIFRA)
New Mexico Department of Agriculture	http://nmdaweb.nmsu.edu	Pesticide enforcement under the Federal Insecticide Fungicide Rodenticide Act (FIFRA)
California Department of Pesticide Regulation (CDPR)	www.epa.gov/Internet	Access OPP's Pesticide Product Information System -- visitors can search this database by company name, chemical name, product name, EPA registration number, and company numbers. Data are updated bi-weekly by OPP staff. Visitors to OPP's website will find a link to the CDPR website.
National Center for Environmental Publications and Information (NCEPI)	www.epa.gov/nceipihom	Publications published by the Environmental Protection Agency may be ordered from this organization.

Agency	World Wide Web Site	Source of Information
Pest Control Technical Magazine	www.pctonline.com	Online magazine covering the pest control industry, business development and pest control information.
National Pest Control Association	www.pestworld.org	Association representing pest management firms worldwide.
Structural Pest Control Commission (NASD)	www.cdc.gov/niosh/nasd/srd/og01 200.html	Advocate and promote, through education, training and enforcement, the safe application of pest control technologies
Pest Control Management Software	http://ourworld.compuserve.com/h omepages/ridenour/pest.htm	Automate inspection processes
Do It Yourself Pest Control	www.doyourownpestcontrol.com	Source of professional pest control supplies for residential and commercial use for pest control.
Pest Management Regulatory Agency	www.hpbl.hwc.ca	Protecting human health and the environment by minimizing risks associated with pest control products.
BioControl Network	www.usit.net/BICONET/	Offers preventive management resources including bio-intensive pest control, organic farm and garden products, educational materials and global community feedback.
National Antimicrobial Information Network	http://ace.orst.edu/info/nain/	Objective information about antimicrobial products. Information is also available on products and published literature.
Northwest Mosquito and Vector Control Association	www.nwmvca.org	Dedicated to the advancement of the mosquito and vector control in the Pacific Northwest.

Agency	World Wide Web Site	Source of Information
COLLEGES AND UNIVERSITIES ONLINE SPECIALIZING IN PESTS AND PEST CONTROL		
Clemson University - Department of Pesticide Regulation	http://dpr.clemson.edu	
Iowa State University - Integrated Pest Management	www.ipm.iastate.edu/ipm	
Simon Fraser University - Master of Pest Management Program	http://mendel.mbb.sfu.ca/mpm/mpm.html	
University of Florida IR-4 Program - Food and Environmental Toxicology Laboratory	http://gnv.ifas.ufl.edu/~foodweb/ir4.htm	
University of Maine - Office of Pest Management, Cooperative Extension Program	http://pmo.umext.maine.edu	
University of Minnesota	www.ent.agri.umn.edu/academic/classes/ipm/ipmsite.htm	
Virginia Tech - Urban Pest Control Center	www.upcrc.vt.edu	
North Carolina State University - Agriculture Institute	www2.ncsu.edu/ncsu/cals/agi/apc/advisors.html	
Ohio State - Laboratory for Pest Control Application Technology	www.oardc.ohio-state.edu/lpcat/	

Appendix C - Illustrations

Illustrations

Chapter 2, Figure "The Trap," pg. 15. Figure supplied by Mr. Chaka Aliim Gale, Clinton, Maryland.

Chapter 5, Figures of insects. Pictures provided by Dr. Paul Morton, Entomology Department. Courtesy of Clemson University, South Carolina.

Chapter 5, Additional figures of insects. Pictures provided by Mr. Brian Fitzek, Public Relations Coordinator. Courtesy of Entomological Society of America, 9301 Annapolis Road, Lanham, MD 20706.

Appendix D - Index

Index

A

G

H

I

J

L

T

W

Y

CAP'S PUBLICATIONS

***From the Garden of Eden to America* by Avaneda D. Hobbs, Ed.D.**
    The function of this book is to provide an in-depth study of the black man, the black church, black leadership, and the social implications of these ingredients within mainstream America. Here is the biblical history of the black man, a prognosis of the black church, and insight into the social and psychological position of the black man about his religion. The book discusses the biblical beginnings of the black man's slavery, the scriptural significance of the Garden of Eden and the history of the races, the origin of the American black church, and a detailed description of all black religious bodies in the United States. Original photos of the denomination's leaders, from the 1700s to the 1900s, are included. Further, this book examines the ingredients required for training effective black leaders in building churches and their role in building an influential church, from an evangelistic point of view. It also provides a comprehensive model to track black churches in any given geographical area in the U.S.

***Who Are We? Building A Knowledge Base About Different Ethnic, Racial, and Cultural Groups in America* by Avaneda D. Hobbs, Ed.D.**
    Experts in diversity agree that the question is whether one's knowledge base can adequately prepare them to lead and gain a competitive edge at their place of influence. This book was prepared to help in presenting one-hour workshops on diversity or cultural awareness. The main goal is to increase one's knowledge base to function in a diverse workforce. It contains self-paced study modules on the different ethnic, racial and cultural groups in America. Each presents the historical background, belief systems and current profiles on the different groups. It provides a step-by-step guide on how to conduct diversity workshops.

***What Pest Control Companies Don't Want You to Know* by Roosevelt McNeil, Jr.**
    Here is an in-depth study of the pest control industry. It elaborates on how some pest control companies rip you off and what they don't tell you that can not only be hazardous to your health, but also to your wallet. This book is designed to provide information to consumers on how to eliminate pests from their home on a permanent basis, how to identify potential problems as a result of unwanted pests, and how to use simple solutions to ward off future problems. Learn the quickest way of killing termites, the best insecticides to get from your local stores, and the best ways of getting rid of mosquitoes. Includes extensive graphics that describe the most common pests and information on providing a pest free environment, as well as a list of reputable pest control companies.

***The New Age Millennium: An Exposé of Symbols, Slogans and Hidden Agendas* by Demond Wilson (formerly of Sanford and Son)**
    This book is a remarkable integration of traditional and contemporary wisdom borne out of the authors' 14 years' rich experience in the worldwide religious arena. The New Age Millennium: An Exposé of Symbols, Slogans and Hidden Agendas is one of the most illuminating books on the New Age movement, one world government, humanism, religion, the Illuminati and Free Masonry. Discover the answers to perplexing questions about where this world is headed to.

ORDER FORM

Type your name, shipping address and telephone number:

Name

Company Name

Shipping Address

City

State/Province   ZIP/Postal Code

Daytime Phone (   )

(In case we have a question about your order)

Please check your product choice here:

DESCRIPTION	QUANTITY	PRICE	TOTAL

Calculate your total cost and indicate method of payment:

Product Cost	$	**Payment method:** (Make checks payable to CAP Publishing & Literary Co.)
US Sales Tax	$	**Check/money order**
Freight	$3.00	**Mastercard**
Total Cost	$	**VISA**
		Credit Card No.

CAP PUBLISHING & LITERARY CO.
P.O. Box 531403
Forestville, MD 20753
e-mail: drvickihay@pobox.com

Expiration Date

Cardholder's Signature